Mathematical Methods for Introductory Physics with Calculus

Third Edition

Mathematical Methods for Introductory Physics with Calculus

Third Edition

Ronald C. Davidson
Professor of Astrophysical Sciences
Princeton University

SAUNDERS GOLDEN SUNBURST SERIES

SAUNDERS COLLEGE PUBLISHING

Harcourt Brace College Publishers
Fort Worth • Philadelphia • San Diego • New York • Orlando • Austin
San Antonio • Toronto • Montreal • London • Sydney • Tokyo

Text Typeface: 10/12 Times Roman
Compositor: Monotype Composition Company, Inc.
Acquisitions Editor: John Vondeling
Developmental Editor: Jennifer Bortel
Managing Editor: Carol Field
Production Service: Elm Street Publishing Services, Inc.
Manager of Art and Design: Carol Bleistine
Art Director: Jennifer Dunn
Cover Designer: Fuiano Design
Director of EDP: Tim Frelick
Sr. Production Manager: Charlene Squibb
Marketing Manager: Marjorie Waldron
Cover Credit: All cover images © FPG International: *Soccer goalie* photographed by Michael Hart; *woman rowing* photographed by Ed Braverman; *skateboarder* photographed by Jeff Divine.

Printed in the United States of America

Mathematical Methods for Introductory Physics with Calculus, Third Edition

ISBN: 0-03-009128-4

Library of Congress Catalog Card Number: 94-5656

4567890123 049 10987654321

Preface

This new edition of *Mathematical Methods for Introductory Physics with Calculus* has evolved from earlier editions co-authored with Professor Jerry B. Marion, now deceased. I would like to dedicate this treatise to the memory of Jerry, a valued friend and colleague, and a preeminent educator.

Students who enter an introductory-level course in physics frequently find themselves unable to keep pace because of a lack of familiarity with the necessary mathematical tools. Such students may experience difficulty in rapidly acquiring the requisite mathematical background because no suitable (and brief) review book can be found. This book is designed to fulfill the need of these students. Assembled here is a short overview of all of the various types of mathematical topics that can reasonably be expected to be needed in an introductory-level course in physics with calculus. The most advanced material in this book is concerned with the basic ideas of differentiation and integration.

An attempt is made to present mathematical ideas from the standpoint of the *physicist;* that is, this book is concerned with *operational* mathematical techniques. Wherever possible, mathematical ideas are introduced with physical examples, and worked problems are frequently included to emphasize each topic. In addition, there are numerous exercises that the student should use to test his or her understanding of each new idea and technique. Answers to these exercises will be found in the back of this book, arranged in random-number fashion (to remove the annoyance of accidentally seeing the answer to the next exercise when checking the last result).

Every student of physics, engineering, or mathematics should possess and know how to use one of the various types of hand-held electronic calculators that are available. To be useful in these areas, a calculator must have, in addition to the normal arithmetic functions, the trigonometric, logarithmic, and exponential functions as well as powers and roots and at least one memory. Calculators that have all of these features are widely available at reasonable prices. If the student is willing to invest somewhat more in a calculator, a unit can be obtained with many additional features such as hyperbolic functions, factorials, rectangular-to-polar conversions, complex number arithmetic, means and standard deviations, linear regression, metric conversions, and so forth. For a still larger investment, programmable calculators are available.

Because most physics courses begin with *mechanics,* all of the examples have been taken from this subject area. Moreover, because the purpose of this book is to provide the student with a *brief* overview of the *mathematical methods* needed in an introductory physics course with calculus, the worked examples do not cover all of the important physics topics that the student is likely to encounter in such a course (e.g., applications of integral calculus to the work-energy theorem, moments of inertia, and other topics have been omitted). While using this book the student should keep in mind that this is *not* a physics text—it is a summary of mathematical methods required in introductory physics and should be used as an adjunct to serious study of the assigned physics text.

Ronald C. Davidson

April 1994

Contents

CHAPTER THREE

Functions and Graphs   69

CHAPTER FOUR

Derivatives   104

CHAPTER FIVE

Vectors   143

CHAPTER SIX

Integral Calculus   179

CHAPTER SEVEN

Experimental Error and Significant Figures   206

APPENDIX A

Some Useful Tables   219

APPENDIX B

Answers to Exercises   227

Index   235

1

Review of Basic Principles

1.1 POWERS OF TEN

On the basis of certain astronomical measurements, we know that the distance from the Sun to the star Alpha Centauri is 40,700,000,000,000,000 meters. And we know that the mass of a single atom of hydrogen is 0.000 000 000 000 000 000 001 67 kilograms. This method of specifying the distance or the mass with a large number of zeroes is awkward and cumbersome. To overcome this difficulty in the expression of large or small numbers, we use a compact notation employing "powers of ten."

Multiplying 10 by itself a number of times, we find

$$10 \times 10 = 100 \qquad\qquad = 10^2$$
$$10 \times 10 \times 10 = 1000 \qquad\qquad = 10^3$$
$$10 \times 10 \times 10 \times 10 \times 10 = 100{,}000 = 10^5$$

The number of times that 10 is multiplied together (that is, the number of *zeroes* that appear in the answer) is used in the result as a superscript to the 10. This superscript is called the *exponent* of 10 or the *power* to which 10 is raised. Also,

$$10^1 = 10$$

and, by convention,

$$10^0 = 1$$

If we express two numbers as powers of ten and then multiply these numbers, we have

$$10^2 \times 10^3 = (10 \times 10) \times (10 \times 10 \times 10) = 10^5 = 10^{(2+3)}$$

The general rule for this kind of operation is

$$10^n \times 10^m = 10^{(n+m)}$$

                     (1.1)

Just as 10^3 means $10 \times 10 \times 10$, we can raise *any* number to a power n by multiplying that number by itself n times. Thus,

$$2^3 = 2 \times 2 \times 2 = 8$$
$$4^2 = 4 \times 4 = 16$$
$$(1.5)^3 = 1.5 \times 1.5 \times 1.5 = 3.375$$

Similarly,

$$(10^2)^3 = (10^2) \times (10^2) \times (10^2)$$
$$= (10 \times 10) \times (10 \times 10) \times (10 \times 10)$$
$$= 10^6$$

The general rule for this kind of operation is

$$(10^n)^m = 10^{(n \times m)}$$

                     (1.2)

EXERCISES

1. $10^7 =$ _____   (Ans. 87)   4. $4^3 =$ _____   (Ans. 21)
2. $100,000,000 =$ _____   (Ans. 75)   5. $(3.1)^4 =$ _____   (Ans. 98)
3. $(10^2) \times (10^7) =$ _____   (Ans. 102)   6. $(1000)^3 =$ _____   (Ans. 115)

1.2   NEGATIVE EXPONENTS

If a power of 10 appears in the denominator of an expression, the exponent is given a negative sign:

$$\frac{1}{10} = 0.1 = 10^{-1}$$

$$\frac{1}{1000} = 0.001 = \frac{1}{10^3} = 10^{-3}$$

In general,

$$\frac{1}{10^m} = 10^{-m} \qquad \text{(1.3)}$$

Multiplying a *positive* power of 10 by a *negative* power of 10 gives

$$10,000 \times 0.01 = 10^4 \times \frac{1}{10^2} = 10^4 \times 10^{-2} = 10^{(4-2)} = 10^2$$

$$\frac{100}{1000} = \frac{10^2}{10^3} = 10^{(2-3)} = 10^{-1} = 0.1$$

In general, combining the rules given in Equations 1.1 and 1.3, we have

$$\frac{10^n}{10^m} = 10^n \times 10^{-m} = 10^{(n-m)} \qquad \text{(1.4)}$$

The same rules for the use of negative exponents apply for numbers other than 10. For example,

$$2^{-2} = \frac{1}{2^2} = \frac{1}{4} = 0.25$$

$$\frac{3^4}{3^2} = 3^{(4-2)} = 3^2 = 9$$

$$\frac{2^3 \times 3^5}{2^2 \times 3^2} = 2^{(3-2)} \times 3^{(5-2)} = 2^1 \times 3^3 = 2 \times 27 = 54$$

EXERCISES

1. $\dfrac{1}{10,000} = $ _____   (Ans. 293)

2. $0.000002 = $ _____   (Ans. 104)

3. $\dfrac{10^7}{10,000} = $ _____   (Ans. 282)

4. $\dfrac{0.01}{10^6} = $ _____   (Ans. 195)

5. $\dfrac{0.2 \times 10^4}{1000 \times 10^{-3}} = $ _____   (Ans. 93)

6. $\dfrac{3^4 \times 5^2}{5 \times 9^2} = $ _____   (Ans. 322)

1.3   PREFIXES

When discussing physical quantities, it frequently proves convenient to use a *prefix* to a unit instead of a power of 10. For example, *centi-* means $\frac{1}{100}$, so *centi*-meter or centimeter (cm) means $\frac{1}{100}$ of a meter (m):

$$1 \text{ m} = 100 \text{ cm} = 10^2 \text{ cm}; \ 1 \text{ cm} = 0.01 \text{ m} = 10^{-2} \text{ m}$$

Similarly, *milli-* means $\frac{1}{1000} = 10^{-3}$ and *mega-* means 10^6:

$$1 \text{ m} = 10^3 \text{ millimeters (mm)}; \ 1 \text{ mm} = 10^{-3} \text{ m}$$

$$\$1,000,000 = 1 \text{ megabuck}$$

Table 1.1 lists some of the most frequently used prefixes.

TABLE 1.1   Prefixes Equivalent to Powers of 10

Prefix	Symbol	Power of 10
giga-	G	10^9*
mega-	M	10^6*
kilo-	k	10^3
centi-	c	10^{-2}
milli-	m	10^{-3}
micro-	μ	10^{-6}
nano-	n	10^{-9}
pico-	p	10^{-12}
femto-	f	10^{-15}

*$10^6 = 1$ *million.* In the United States, $10^9 = 1$ *billion,* but the European convention is that $10^9 = 1000$ million and that 1 billion $= 10^{12}$; the prefix *giga-* is internationally agreed on to represent 10^9.

EXERCISES

1. 1 microsecond (μs) = _____ s   (Ans. 146)
2. 1 kilometer (km) = _____ cm   (Ans. 182)
3. 1 cm = _____ m   (Ans. 91)
4. (1 km) $\times$ (1 m) = _____ m^2 (Ans. 310)

1.4   CALCULATIONS WITH POWERS OF TEN

By using powers of 10, many types of calculations are made considerably easier. First, we note that *any* number can be expressed in terms of a power of 10 by writing, for example,

$$6400 = 6.4 \times 10^3$$

$$0.0137 = 1.37 \times 10^{-2}$$

$$970,000 = 0.97 \times 10^6$$

Notice that in changing from an ordinary number to a number expressed as a power of 10, the exponent of 10 corresponds to the number of places that the decimal has been moved. (A positive exponent means that the decimal has been moved to the left and a negative exponent means that the decimal has been moved to the right.) Thus, in expressing 6400 in terms of a power of 10, we move the decimal three places to the left to obtain 6.4×10^3. And in expressing 0.0137 in terms of a power of 10, we move the decimal two places to the right, obtaining 1.37×10^{-2}.

Usually, we express a quantity in this notation by writing the multiplying factor as a number between 0.1 and 10. That is, we write $240,000 = 2.4 \times 10^5$ instead of 240×10^3. (See, however, the discussion in Section 7.3.)

$$\frac{42,000,000}{3000} = \frac{4.2 \times 10^7}{3 \times 10^3} = \frac{4.2}{3} \times \frac{10^7}{10^3} = 1.4 \times 10^4$$

$$0.0012 \times 0.000\,003 = (1.2 \times 10^{-3}) \times (3 \times 10^{-6})$$

$$= (1.2 \times 3) \times (10^{-3} \times 10^{-6}) = 3.6 \times 10^{-9}$$

EXERCISES

1. Express the distance from the Sun to Alpha Centauri (in m) as a power of 10. (See Section 1.1.)

   (Ans. 183)

2. Express the mass of a hydrogen atom (in kilograms) as a power of 10. (See Section 1.1.)   (Ans. 250)

3. $\dfrac{480,000}{8000} = $ _____   (Ans. 232)

4. $1600 \times 0.003 = $ _____   (Ans. 355)

5. $\dfrac{18,000,000}{0.0003} = $ _____   (Ans. 101)

6. $\dfrac{320 \times 0.0048}{0.32 \times 2400} = $ _____   (Ans. 323)

7. $0.09 \times 0.4 \times 4000 = $ _____   (Ans. 110)

1.5   THINKING IN ORDERS OF MAGNITUDE

One power of 10 is sometimes referred to as an *order of magnitude*. Thus, "a dollar is an order of magnitude more valuable than a dime and two orders of magnitude (a factor of 100) more valuable than a penny." More frequently, the term *order of magnitude* is applied in an approximate sense. Thus, "the Earth is two orders of magnitude more massive than the Moon." (Actually, the mass of the Earth is 81 times the mass of the Moon.) Or, since the meter is 100 centimeters and the yard is 91.44 centimeters, we say "the meter and the yard are of the same order of magnitude."

Because we deal so frequently with physical quantities that are very large or very small, it is extremely helpful if we cultivate the habit of thinking about and referring to these quantities in orders of magnitude. Even if we do not know the precise value of a certain quantity it is still useful to know the order of magnitude of this size. For example, the diameter of an atom is of the order of 10^{-8} cm. If we perform some calculation and obtain a result that says that the size of an atom is 10^{-5} cm, then we know immediately that something is wrong in the calculation!

We will use the symbol $\sim$ to indicate "is of the order of magnitude of." Often we include one number in addition to the power of 10, so that a somewhat more precise (but still approximate) value results. Some typical order-of-magnitude quantities are as follows:

(a)   There are $\sim 3 \times 10^7$ seconds in a year.
(b)   Light travels ~ 1 foot in 10^{-9} seconds.
(c)   The diameter of a nucleus is $\sim 10^{-4}$ times the diameter of an atom.
(d)   The mass of the Sun is $\sim 3 \times 10^5$ times the mass of the Earth.
(e)   The mass of a hydrogen atom is $\sim 2 \times 10^3$ times the mass of an electron.

None of the values given above is *precise,* but an appreciation of the *approximate* value of a quantity permits the rapid evaluation of a situation and can often be used to determine whether a particular calculation is reliable.

EXERCISES   *Give order-of-magnitude answers for the following:*

1. The Universe is thought to be $\sim 3 \times 10^{10}$ years old. Express this age in seconds.   (Ans. 351)
2. The mass of the Sun is about 10^{30} kg and consists mainly of hydrogen. Approximately how many atoms of hydrogen are there in the Sun? (Refer to the mass of the hydrogen atom given in Section 1.1.)
     (Ans. 227)

3. The area of the United States (including Alaska and Hawaii) is $\sim 3,600,000$ square miles and the population is $\sim 230,000,000$. What is the population density (persons per square mile)?   (Ans. 199)
4. Express your age in seconds.

1.6   FRACTIONAL EXPONENTS

An *integer* exponent represents the *power* to which a number is raised; for example, $10^3 = 10 \times 10 \times 10 = 1000$ or $3^2 = 3 \times 3 = 9$. *Fractional* exponents also have meaning and these exponents are manipulated according to the same rules as for integer exponents. For example,

$$4^{1/2} \times 4^{1/2} = 4^{(1/2 + 1/2)} = 4^1 = 4, \text{ or } (4^{1/2})^2 = 4$$

$$27^{1/3} \times 27^{1/3} \times 27^{1/3} = 27^{(1/3 + 1/3 + 1/3)} = 27^1 = 27, \text{ or } (27^{1/3})^3 = 27$$

What is the meaning of a fractional exponent? In the examples above, notice that if a number $N^{1/m}$ is raised to the power m, the result is N itself. We call $N^{1/m}$ the mth *root* of N, and we also use the equivalent notation,

$$\sqrt[m]{N} = N^{1/m} \tag{1.5}$$

If $m = 2$, this factor is usually suppressed from the radical sign, $\sqrt{}$. For example,

$$4 \times 4 = 4^2 = 16, \text{ so that } \sqrt{16} = 16^{1/2} = 4$$

$$\sqrt{5^2} = (5^2)^{1/2} = 5^{(2 \times 1/2)} = 5^1 = 5$$

$$2 \times 2 \times 2 = 2^3 = 8, \text{ so that } \sqrt[3]{8} = 8^{1/3} = 2$$

This latter result can also be obtained by writing

$$\sqrt[3]{8} = \sqrt[3]{2^3} = (2^3)^{1/3} = 2^{(3 \times 1/3)} = 2^1 = 2$$

More complicated fractional exponents are handled in the following way:

$$4^{3/2} = 4^{(1/2 \times 3)} = (4^{1/2})^3 = (2)^3 = 8$$

or,

$$4^{3/2} = 4^{(3 \times 1/2)} = (4^3)^{1/2} = (64)^{1/2} = \sqrt{64} = 8$$

Notice that the *numerator* of the fractional exponent always represents a *power* to which the number is raised and the *denominator* always represents the *root* of the number.

$$\sqrt[3]{8000} = (8000)^{1/3} = (2^3 \times 10^3)^{1/3} = 2 \times 10 = 20$$

$$(2.7 \times 10^4)^{1/3} = (27,000)^{1/3} = (27 \times 10^3)^{1/3} = \sqrt[3]{27} \times 10 = 3 \times 10 = 30$$

$$(64,000)^{2/3} = (64 \times 10^3)^{2/3} = (\sqrt[3]{64})^2 \times 10^{(3 \times 2/3)} = 4^2 \times 10^2 = 1600$$

In writing equations such as $\sqrt{4} = 2$, we are not being properly general. The equation $\sqrt{4} = 2$ means that $2^2 = 2 \times 2 = 4$. But it is also true that $(-2) \times (-2) = 4$, so we could equally well write $\sqrt{4} = -2$. There is always an ambiguity of sign in extracting roots. We will usually want the *positive* sign (the *physics* of the problem will dictate which sign is physically meaningful), but we should properly indicate the possibility of either sign by writing $\sqrt{4} = \pm 2$. (The symbol $\pm$ means ''positive or negative.'')

EXERCISES

1. $36^{1/2} =$ _____   (Ans. 109)   **5.** $\left(\dfrac{81 \times 10^7}{3^2 \times 10^5}\right)^{1/2}$ (Ans. 97)

2. $25^{3/2} =$ _____   (Ans. 155)

3. $81^{1/4} =$ _____   (Ans. 262)   **6.** $\left[\left(\dfrac{1.296 \times 10^5}{400}\right)^{1/2} + 9\right]^{1/3} =$ _____   (Ans. 20)

4. $64^{1/3} =$ _____   (Ans. 94)

1.7   MATHEMATICAL NOTATION

In ordinary equations we use the symbol $=$ to denote equality of two quantities:

$$y = 16.27t^2 \quad \text{or} \quad A \times B = C$$

Even if we did not know the factor 16.27 that occurs in the above equation, we could still state that y is *proportional* to t^2, and we would write

$$y = kt^2 \quad \text{or} \quad y \propto t^2$$

Or, if we knew only that the factor is *approximately* equal to 16, we would write

$$y \cong 16t^2$$

We have already seen that the symbol $\sim$ (or $\approx$) is used to indicate that two quantities are *very approximately* equal or have the same *order of magnitude.*

The symbols $<$ and $>$ mean, respectively, *is less than* and *is greater than*; for example,

$$\text{area of Canada} > \text{area of Argentina}$$

$$\text{mass of the Earth} < \text{mass of Jupiter}$$

If a quantity *is very much smaller* or *very much larger* than another quantity, we use a double symbol:

$$\text{area of Canada} \gg \text{area of Luxembourg}$$

$$\text{mass of the Earth} \ll \text{mass of the Milky Way Galaxy}$$

Sometimes we know only that a quantity is smaller or larger than some poorly defined quantity. In such a case we use the symbol $\lesssim$ (or $\gtrsim$) to denote *is less than about* (or *is greater than about*). For example,

$$\text{Population of the United States} \gtrsim 2 \times 10^8$$

We frequently find it convenient to use a shorthand notation to indicate the *change* in the value of a quantity. If an object is located at the position $x_1 = 2$ cm at a certain time and if at a later time the location is $x_2 = 9$ cm, we say that the distance moved (or the change in x) is $x_2 - x_1 = 9$ cm $- 2$ cm $= 7$ cm. That is, we take the *final* position (x_2) and subtract from it the *initial* position (x_1). This change in x is often denoted by the symbol Δx:

$$\Delta x = x_{\text{final}} - x_{\text{initial}} = x_2 - x_1 \tag{1.6}$$

The symbol Δx does *not* imply the product of Δ and x, but means "the change in x" or "an increment of x." In general, a Greek delta, Δ, in front of a quantity means the *change* in that quantity; e.g., $t_2 - t_1 = \Delta t =$ time difference. Δt can be either *positive* ($\Delta t > 0$) or *negative* ($\Delta t < 0$), depending on whether t_2 is greater or smaller than t_1.

FIGURE 1.1  $\Delta x = x_2 - x_1$.

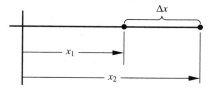

Sometimes we are interested only in the *magnitude* of a quantity and not whether it carries a positive or a negative sign. We denote the magnitude of a quantity by vertical bars, $|x|$. That is, for positive values of x, $|+x| = x$ and $|-x| = x$.

$$|3x - 7x| = |-4x| = 4x$$

In many types of physical problems we find it necessary to add a series of related numbers. For example, in order to determine the total mass M of a system composed of many individual masses, $m_1, m_2, m_3, m_4, \ldots$, we would write

$$M = m_1 + m_2 + m_3 + m_4 + \cdots + m_N \tag{1.7}$$

where N represents the total number of particles in the system. Because this situation arises so frequently, a special notation has been devised to abbreviate lengthy summations. In order to represent Equation 1.7, we write

$$M = \sum_{i=1}^{N} m_i \tag{1.8}$$

Here, we have written a typical particle mass as m_i and we allow i (which is called the *summation index*) to take on the integer numbers from 1 to N. That is, m_i (for $i = 1$) $= m_1$, m_i (for $i = 2$) $= m_2$, etc. The symbol Σ (which is called the *summation symbol*) means "sum all of the quantities which follow." Beneath Σ is the indication of the index to be summed (here, i) and the *starting* value of the index (here, 1); above the Σ is the *terminating* value of the index (here, N).

TABLE 1.2   Mathematical Symbols and Their Meanings

Symbol	Meaning
$=$	is equal to
$\propto$	is proportional to
$\cong$	is approximately equal to
$\approx$ or $\sim$	is *very* approximately equal to; is of the order of magnitude of
$>$ $(<)$	is greater (less) than
$\gg$ $(\ll)$	is much greater (less) than
$\gtrsim$ $(\lesssim)$	is greater (less) than about
Δx	change in x
$\lvert x \rvert$	magnitude of x
$\displaystyle\sum_{i=1}^{N} m_i$	sum of m_i from $i = 1$ to $i = N$
$\displaystyle\sum_{i} m_i$	sum all m_i for all physically allowed values of i

If we sum the integers from 1 to N, the result is $\frac{1}{2} N(N + 1)$. In the notation of Equation (1.8), this is expressed as

$$\sum_{i=1}^{N} i = \frac{1}{2} N(N + 1) \tag{1.9}$$

(Try this for a few values of N.) Or, if we sum the *squares* of the integers from 1 to N, we find

$$\sum_{m=1}^{N} m^2 = \frac{N}{6} (2N^2 + 3N + 1) \tag{1.10}$$

Here, we have used a summation index of m instead of i; it does not matter at all what letter or symbol we use for the index because it does not appear in the final result.

The meanings of the symbols we will find useful in this book are summarized in Table 1.2.

EXERCISES   *Insert the appropriate symbol between the following pairs of quantities (in some cases there may be two appropriate symbols—give both):*

1. Area of Soviet Union   area of United States
(Ans. 143)

2. Height of Mt. Whitney   Mt. Everest   (Ans. 118)

3. Gross National Product 10^6 megabucks
(Ans. 357)

4. If $\dfrac{x}{y} = 10^{-5}$, then x y (Ans. 308)

5. Mass of atom   mass of apple   (Ans. 327)

6. Write out the following sum

$$\sum_{p=1}^{N} (2p - 1)$$

for $N = 3$ and for $N = 5$. Verify that the sum is equal to N^2.

7. Write out the first few terms in the sum

$$\sum_{n=0}^{N} \frac{(-1)^{2n}}{n + 1}$$ (Ans. 99)

1.8   FUNDAMENTAL PHYSICAL QUANTITIES

Physical quantities not only have *magnitudes,* specified by *numbers,* but also *dimensions* or *units.* It makes no sense to say that a certain length is ''75'' unless we also state the appropriate units—feet, meters, miles, or whatever. Although we encounter a wide variety of physical quantities that require units for their complete specification—for example, distance, force, energy, momentum, and electric field strength—these various units can be expressed in terms of only *three* fundamental quantities. The basic units of physical measure are those of *length, time,* and *mass*—the dimensions of all other mechanical quantities can be expressed in terms of the units of these three. For example, *speed* is measured by the distance traveled in a certain time and so the dimensions of speed are *length/time.* In the next section we will review the dimensions of other physical quantities.

The system of units most widely used in science is the *metric system.* It has become standard practice to use the particular metric units and symbols that have been prescribed by the commission on ''Le Système International d'Unités.'' We refer to these as *SI units.* In the SI system, the base units of mechanics are the *meter* (length), the *kilogram* (mass), and the *second* (time). (When electric and thermodynamic problems are considered, it is necessary to introduce additional base units.) Also in use at the present time is the *British engineering system,* in which the standard units are the *yard* (length), the *pound-mass* (mass), and the *second* (time). A few parts of the world (most notably the United States) have not yet converted to the metric system, but it seems inevitable that the metric system will replace the British system.

The fundamental units of measure in the metric and British systems are summarized in Table 1.3.

TABLE 1.3   Fundamental Units of Measure

	Metric (SI)	British
Length	Meter (m) 1 m = 100 cm	Yard (yd) 1 yd = 36 in 1 in = 2.54 cm
Mass	Kilogram (kg) 1 kg = 1000 g	Pound-mass (lb) 1 lb = 0.4536 kg
Time	Second (s)	Second (s)

If we wish to convert a certain length from one system of units to another, we use the following procedure. Since 1 in = 2.54 cm, we can form the ratios

$$\frac{2.54 \text{ cm}}{1 \text{ in}} = 1, \quad \text{or} \quad \frac{1 \text{ in}}{2.54 \text{ cm}} = 1$$

These ratios are just *unity,* so we can multiply any quantity by either ratio without changing the value. Thus,

$$15 \text{ in} = (15 \text{ in}) \times \left(\frac{2.54 \text{ cm}}{1 \text{ in}}\right) = 38.10 \frac{\text{cm-in}}{\text{in}} = 38.10 \text{ cm}$$

where the *inches* in the numerator and denominator cancel. Also,

$$48 \text{ cm} = (48 \text{ cm}) \times \left(\frac{1 \text{ in}}{2.54 \text{ cm}}\right) = 18.9 \text{ in}$$

$$36 \text{ kg-m} = (36 \text{ kg-m}) \times \left(\frac{10^3 \text{ g}}{1 \text{ kg}}\right) \times \left(\frac{10^2 \text{ cm}}{1 \text{ m}}\right) = 3.6 \times 10^6 \text{ g-cm}$$

$$60 \text{ mi/h} = (60 \text{ mi/h}) \times \left(\frac{5280 \text{ ft}}{1 \text{ mi}}\right) \times \left(\frac{1 \text{ h}}{3600 \text{ s}}\right) = 88 \text{ ft/s}$$

We will rarely use British units except in a few elementary examples involving the motion of objects.

Any quantity that does not have units is said to be *dimensionless* or a *pure number.* For example, the *ratio* of two physical quantities with the same dimensions is *dimensionless:*

$$R = \frac{8 \text{ cm}}{4 \text{ cm}} = 2$$

The quantity R is a *pure number.*

EXERCISES

1. 10 years = _____ s   (Ans. 274)
2. 504 km/h = _____ m/s   (Ans. 242)
3. 1 km = _____ cm   (Ans. 121)
4. 6 lb = _____ g   (Ans. 312)

5. 6.6 m/s = _____ yd/s   (Ans. 277)
6. 10 in = _____ mm   (Ans. 162)
7. 123 kg-m = _____ lb-in   (Ans. 33)
8. 1 yd = _____ cm   (Ans. 244)

1.9   DENSITY

The fundamental physical quantities—length, mass, and time—must frequently be used in combinations to describe certain physical concepts or properties. One of these properties that is easy to understand is *density,* the ratio of an object's mass to its volume:

$$\text{density} = \frac{\text{mass}}{\text{volume}}$$

(1.11)

or, in symbols,

$$\rho = \frac{M}{V}$$

(1.12)

The density is an *intrinsic* property of a substance and does not depend on its shape or its volume. The SI unit of density is the kg/m^3, but we frequently see densities expressed in g/cm^3 (because the density of water is just 1 g/cm^3). The densities of some common substances are given in Table 1.4.

TABLE 1.4 Densities of Some Common Substances

Substance	Density kg/m^3	Density g/cm^3
Air (normal conditions*)	1.3	1.3×10^{-3}
Water	1.0×10^3	1.0
Aluminum	2.7×10^3	2.7
Iron	7.9×10^3	7.9
Lead	1.13×10^4	11.3
Gold	1.93×10^4	19.3

*Temperature = 0°C and pressure = 1 atmosphere.

▶ **EXAMPLE 1.9.1** Lead bricks (used for shielding radioactive materials) commonly have dimensions of 2 in × 4 in × 8 in. What is the mass of such a brick?

$$\text{Volume} = V = (2 \text{ in}) \times (4 \text{ in}) \times (8 \text{ in}) = 64 \text{ in}^3$$

$$= (64 \text{ in}^3) \times \left(\frac{2.54 \text{ cm}}{1 \text{ in}}\right)^3$$

$$= (64 \text{ in}^3) \times \left(\frac{16.39 \text{ cm}^3}{1 \text{ in}^3}\right)$$

$$= 1049 \text{ cm}^3$$

(1)

Using $\rho = 11.3$ g/cm^3 from Table 1.4 we have

$$M = \rho V = \left(\frac{11.3 \text{ g}}{\text{cm}^3}\right) \times (1049 \text{ cm}^3)$$

$$= 1.185 \times 10^4 \text{ g}$$

$$= 11.85 \text{ kg} \qquad\qquad\qquad \textbf{(2)}$$

or, in British units,

$$M = (11.85 \text{ kg}) \times \left(\frac{1 \text{ lb}}{0.4536 \text{ kg}}\right)$$

$$= 26.12 \text{ lb}$$

The rule-of-thumb is that the mass of a standard lead brick is $\sim$25 lb or $\sim$12 kg.

▶ **EXAMPLE 1.9.2**   What is the average density of the Earth?

The radius of the Earth is $R_E = 6.38 \times 10^6$ m, so that

$$V = \frac{4}{3} \pi R_E^3 = \frac{4}{3} \pi (6.38 \times 10^6 \text{ m})^3$$

$$= 1.09 \times 10^{21} \text{ m}^3$$

The mass of the Earth is $M_E = 5.98 \times 10^{24}$ kg; thus,

$$\rho = \frac{M_E}{V} = \frac{5.98 \times 10^{24} \text{ kg}}{1.09 \times 10^{21} \text{ m}^3} = 5.49 \times 10^3 \text{ kg/m}^3 = 5.49 \text{ g/cm}^3$$

This is the density averaged throughout the entire Earth. Actually, the core has a considerably higher density ($\sim$12 g/cm^3) and the mantle (the rocky material near the surface) has a density of $\sim$3 g/cm^3.

EXERCISES

1. What is the mass of a cubic meter of lead?
   (Ans. 270)

2. What is the mass of a cube of gold with sides of length 3 cm?
   (Ans. 74)

3. A gold sphere has a radius of 5 cm. What is its mass?
   (Ans. 191)

4. A pipe with a diameter of 20 cm contains a column of water 5 m high. What is the mass of the water?
   (Ans. 130)

5. What is the density (the *average* density) of the Sun? (Radius = 6.96×10^8 m, mass = 1.99×10^{30} kg.)

(Ans. 71)

6. What is the difference in mass between two spheres (each with radius 10 cm), one of which is lead and one of which is gold? (Ans. 92)

1.10 THE DIMENSIONS OF PHYSICAL QUANTITIES

When we present physical examples of mathematical ideas in later chapters, we will have occasion to use a variety of physical concepts. Therefore, we summarize here the important physical quantities that appear in the area of *mechanics*. Even though the dimensions of each of these quantities can be expressed in terms of length, mass, and time, it is customary to attach special names to the units for certain quantities. We refer to these as *derived* units.

Velocity (v). Speed, or the rate at which position changes with time:*

$$v = \frac{\Delta x}{\Delta t}, \quad \text{m/s, cm/s, etc.} \tag{1.13}$$

Acceleration (a). Speeding up or slowing down; the rate at which velocity changes with time:

$$a = \frac{\Delta v}{\Delta t}, \quad \text{(m/s)/s or m/s}^2, \text{ etc.} \tag{1.14}$$

Momentum (p). The product of a body's mass and its velocity:

$$p = mv, \quad \text{kg-m/s} \tag{1.15}$$

Force (F). A push or a pull; the fundamental definition is in terms of the rate at which a body's momentum changes with time:

$$F = \frac{\Delta p}{\Delta t}, \quad \text{kg-m/s}^2 \tag{1.16}$$

Another important equation for force is Newton's second law:

$$F = ma, \quad \text{kg-m/s}^2 \tag{1.17}$$

---

*The terms *velocity* and *speed* are similar in meaning, but as used in physics there is an important distinction. The concept of motion involves not only the *rate* at which an object moves but also the *direction* of movement. The term *velocity* is usually reserved for use when the direction of the motion is important (that is, when the *vector* character of velocity is important; see Chapter 5). The term *speed* is used when only the rate of movement (regardless of direction) is to be specified.

The unit of force in the SI system is given a special name:

$$1 \text{ kg-m/s}^2 = 1 \text{ newton (N)} \qquad (1.18)$$

Work (W). The expenditure of energy; the product of force (F) and the distance (s) through which the force acts:

$$W = Fs, \quad \text{kg-m}^2/\text{s}^2 \qquad (1.19)$$

The unit of work in the SI system is given a special name:

$$1 \text{ kg-m}^2/\text{s}^2 = 1 \text{ N-m} = 1 \text{ joule (J)} \qquad (1.20)$$

Energy (E). The capacity to do work; energy comes in many forms—in mechanical situations we identify *kinetic* energy (the energy possessed by a body because of its *motion*) and *potential* energy (the energy possessed by a body because of its *position*):

$$\text{Kinetic energy: } E_K = \frac{1}{2}mv^2, \quad \text{kg-m}^2/\text{s}^2, \text{ J} \qquad (1.21)$$

For the particular case of a body of mass m in the gravitational field near the surface of the Earth, the potential energy is

$$\text{Potential energy: } E_P = mgh, \quad \text{kg-m}^2/\text{s}^2, \text{ J} \qquad (1.22)$$

where g is the acceleration due to gravity (i.e., the acceleration experienced by a freely falling body) and where h is the height of the body above some reference position (e.g., the surface of the Earth). Near the surface of the Earth, g has the value

$$g \cong 9.8 \text{ m/s}^2$$
$$\cong 32 \text{ ft/s}^2 \qquad (1.23)$$

Notice that *work* and *energy* have the *same* dimensions. Table 1.5 summarizes the units of these various mechanical quantities.

TABLE 1.5 Some Physical Quantities and Their Units

Quantity	Equation(s)	SI Units
Velocity (or speed*)	$v = \dfrac{\Delta x}{\Delta t}$	m/s
Acceleration	$a = \dfrac{\Delta v}{\Delta t}$	m/s²
Momentum	$p = mv$	kg-m/s
Force	$F = \dfrac{\Delta p}{\Delta t}$ $F = ma$	$\left\{ \begin{array}{l} \text{kg-m/s}^2 \\[4pt] = \text{newton (N)} \end{array} \right.$
Work	$W = Fs$	$\left\{ \begin{array}{l} \text{kg-m}^2/\text{s}^2 \end{array} \right.$
Energy	$E_K = \dfrac{1}{2} mv^2$ $E_P = mgh$	$\left\{ \begin{array}{l} = \text{N-m} \\[4pt] = \text{joule (J)} \end{array} \right.$

*See footnote on p. 15.

▶ **EXAMPLE 1.10.1**   A particle has a velocity of 20 ft/s at $t_1 = 4.5$ s and a velocity of 48 ft/s at $t_2 = 6.8$ s. Express the acceleration in m/s²:

First, we convert the velocities to m/s:

$$v_1 = 20 \text{ ft/s} = \left(20 \frac{\text{ft}}{\text{s}} \right) \times \left(\frac{12 \text{ in}}{1 \text{ ft}} \right) \times \left(\frac{2.54 \text{ cm}}{1 \text{ in}} \right) \times \left(\frac{1 \text{ m}}{100 \text{ cm}} \right)$$

$$= 6.1 \text{ m/s}$$

$$v_2 = 48 \text{ ft/s} = \left(48 \frac{\text{ft}}{\text{s}} \right) \times \left(\frac{12 \text{ in}}{1 \text{ ft}} \right) \times \left(\frac{2.54 \text{ cm}}{1 \text{ in}} \right) \times \left(\frac{1 \text{ m}}{100 \text{ cm}} \right)$$

$$= 14.6 \text{ m/s}$$

Then,

$$a = \frac{\Delta v}{\Delta t} = \frac{v_2 - v_1}{t_2 - t_1} = \frac{14.6 \text{ m/s} - 6.1 \text{ m/s}}{6.8 \text{ s} - 4.5 \text{ s}}$$

$$= \frac{8.5 \text{ m/s}}{2.3 \text{ s}} = 3.7 \text{ m/s}^2$$

The sign of a is *positive,* so the particle was *speeding up;* if v_1 had exceeded v_2, the sign of a would have been *negative,* indicating that the particle was *slowing down.*

▶ **EXAMPLE 1.10.2**   What is the kinetic energy (in joules) of a 0.80-kg particle that is moving with a velocity of 9.0 m/s?

$$E_K = \frac{1}{2}mv^2 = \frac{1}{2} \times (0.8 \text{ kg}) \times (9 \text{ m/s})^2$$

$$= 36 \text{ J}$$

▶ **EXAMPLE 1.10.3**   What force is necessary to increase the velocity of a 2-kg ball from 2 m/s to 6 m/s in 0.5 s?

The momentum change is

$$\Delta p = m\Delta v = (2 \text{ kg}) \times (6 \text{ m/s} - 2 \text{ m/s})$$

$$= 8 \text{ kg-m/s}$$

Then,

$$F = \frac{\Delta p}{\Delta t} = \frac{8 \text{ kg-m/s}}{0.5 \text{ s}} = 16 \text{ N}$$

▶ **EXAMPLE 1.10.4**   A 5-kg block is raised to a height of 12 m and given a velocity of 15 m/s. What is the energy of the block?

The energy consists of two parts:

$$E = E_P + E_K = mgh + \frac{1}{2}mv^2$$

$$= (5 \text{ kg}) \times (9.8 \text{ m/s}^2) \times (12 \text{ m}) + \frac{1}{2}(5 \text{ kg}) \times (15 \text{ m/s})^2$$

$$= 588 \text{ kg-m/s}^2 + 562.5 \text{ kg-m/s}^2$$

$$= 1150.5 \text{ J}$$

EXERCISES

1. Newton's law of universal gravitation states that the gravitational force between two bodies with masses m_1 and m_2, which are separated by a distance r, is $F = Gm_1m_2/r^2$, where G is a constant. What are the dimensions of G in SI units?   (Ans. 57)

2. A constant force of 5 N acts on a body and the body moves a distance of 3 m. What amount of work was done?   (Ans. 141)

3. A 5-kg object is moved from the surface of the Earth to a height of 50 m. What is the change in potential energy of the object? (Ans. 35)

4. What force (in *newtons*) is required to change the momentum of a body from 50 kg-m/s to 80 kg-m/s in a period of 3 s? (Ans. 22)

5. A 2-g object is moving with a velocity of 4 cm/s. What is the kinetic energy of the object? (Ans. 56)

1.11 SIMPLE ALGEBRAIC EQUATIONS

The most elementary type of algebraic equation has the form

$$ax + b = 0 \tag{1.24}$$

where a and b are numbers and where x represents the unknown quantity. Equations of this type are called linear equations because they involve only the *first power* of x and, when plotted on a graph, they appear as *straight lines* (see Section 3.3).

Equation (1.24) is solved in two steps. First, transpose b to the right-hand side of the equation. That is, remove $+b$ from the left-hand side and place $-b$ on the right-hand side. (This is the same as subtracting b from both sides of the equation, an operation that is always allowed.) Next, divide both sides of the equation by a:

$$ax + b = 0$$
$$ax = -b$$
$$x = -\frac{b}{a}$$

Both steps in solving the simple linear equation are carried out by performing the *same* operation on *both* sides of the equation. Any linear equation, even though it may appear more complex than Equation (1.24), can always be manipulated into this form and solved in this way.

► **EXAMPLE 1.11.1** Solve $\dfrac{1}{a - x} = b$

First, multiply both sides by $(a - x)$:

$$\frac{a - x}{a - x} = b(a - x)$$
$$1 = ab - bx$$

Next, transpose:

$$bx = ab - 1$$

Finally, divide by b:

$$x = \frac{ab - 1}{b} = a - \frac{1}{b}$$

The solution can be checked by substituting back into the original equation to verify that it yields $1 = 1$. The results of all algebraic computations should be checked this way.

EXERCISES   *In each case, solve for x:*

1. $6x + 5 = 15 + x$ (Ans. 235)   **4.** $\dfrac{a + b}{x} = c$ (Ans. 81)

2. $18x + a - b = 9 + 12x$ (Ans. 247)

3. $\frac{1}{7} x + 5 = \frac{2}{3} x - 6$ (Ans. 126)   **5.** $\dfrac{4}{x - 4} = \dfrac{7}{x - 5}$ (Ans. 103)

1.12 SIMULTANEOUS LINEAR EQUATIONS

In certain types of physical problems we find that we have an equation involving more than a single unknown. For example,

$$6x + 2y = 6$$

Such an equation does not yield a unique solution; it is easy to verify that $(x = 1, y = 0)$, $(x = 4, y = -9)$, and $(x = \frac{1}{2}, y = \frac{3}{2})$ are all solutions of the above equation.

In order to have a unique solution for a problem involving *two* unknowns, we must have *two* equations. In general, if we have a problem with n unknown quantities, we need n equations in order to obtain a solution.

Two methods of solving a pair of simple linear equations are given in the following example.

▶ **EXAMPLE 1.12.1**   Consider the pair of equations,

$$6x + 2y = 6 \qquad (1)$$

$$8x - 4y = 28 \qquad (2)$$

The solution can be obtained by first solving one of the equations for y and then substituting this expression into the other equation. The result will be a single equation in the unknown x and the solution can be obtained by one of the methods described in the preceding section:

Solving (1) for y,

$$2y = 6 - 6x$$

$$y = 3 - 3x \tag{3}$$

Substituting (3) into (2), we find

$$8x - 4(3 - 3x) = 28$$

$$8x - 12 + 12x = 28$$

$$20x = 40$$

$$x = 2 \tag{4}$$

Finally, substituting (4) into (1), we obtain

$$6 \cdot 2 + 2y = 6$$

$$2y = -6$$

$$y = -3 \tag{5}$$

Therefore, the solution is $(x = 2, y = -3)$.

Alternatively, we can solve these equations by multiplying (1) by 2 and then adding to (2):

$$2 \cdot (6x + 2y) = 2 \cdot 6$$

$$12x + 4y = 12 \tag{6}$$

Then, adding (2) and (6),

$$\begin{array}{r} 8x - 4y = 28 \\ \underline{12x + 4y = 12} \\ 20x \quad\quad = 40 \end{array}$$

$$x = 2 \tag{7}$$

And substituting (7) into (1) again produces $y = -3$.

Notice that we choose to multiply (1) by 2 (and not by some other number) in order to make the term involving y equal in magnitude to the corresponding term in (2). Then, when the two equations are added, the y-term cancels. In some cases, we must multiply each equation by a different factor in order to ensure that one of the terms will cancel. Thus, we could have multiplied (1) by 4 and (2) by 3 and then *subtracted* the resulting equations to eliminate the x-term. (Work out the example in this way.)

▶ **EXAMPLE 1.12.2**   The position x of a particle at time t moving along a straight line in the absence of a force is given by

$$x = x_0 + v_0 t$$

where x_0 is the position at time $t = 0$, v_0 is the velocity at time $t = 0$, and the acceleration $a = F_0/m$ is taken to be zero.

Suppose that two particles move along the straight line, the first particle beginning at position $x_0 = 0$ with initial velocity $v_0 = 5$ m/s, and the second particle beginning further to the right at position $x_0 = 2$m with smaller initial velocity $v_0 = 2$ m/s. At what position x and at what time t will the first particle overtake ("collide" with) the second particle?

The motion of the first particle satisfies

$$x = 0 + 5t$$

whereas the motion of the second particle satisfies

$$x = 2 + 2t$$

To determine the position x and time t at which the two particles "collide," we solve these two equations simultaneously for x and t. Substituting $x = 5t$ into $x = 2 + 2t$, we obtain

$$5t = 2 + 2t$$
$$5t - 2t = 2$$
$$3t = 2$$
$$t = \frac{2}{3} \text{ s}$$

Substituting $t = \frac{2}{3}$ s into $x = 5t$, we obtain

$$x = 5 \cdot \frac{2}{3}$$
$$= \frac{10}{3}$$
$$= 3\frac{1}{3} \text{ m}$$

Therefore, the faster moving first particle overtakes ("collides" with) the second particle at position $x = 3\frac{1}{3}$ m after a time $t = \frac{2}{3}$ s has elapsed.

EXERCISES

Solve the following pairs of equations where x and y are in meters and t is in seconds:

1. $x + 2y = 8$
    $x - y = 4$

(Ans. 59)

2. $18x + y = 5$
    $12x - y = 25$

(Ans. 295)

3. $2x - 3y = 7$
    $3x + 2y = 2$

(Ans. 186)

1.13   QUADRATIC EQUATIONS

Quadratic equations involve the *second* power of the unknown quantity and can be reduced to the form

$$ax^2 + bx + c = 0 \qquad \text{(1.25)}$$

The solution of this equation is

$$x = \frac{-b \pm \sqrt{b^2 - 4ac}}{2a} \qquad \text{(1.26)}$$

Focus on the term $\sqrt{b^2 - 4ac}$. We have three cases to consider:
(1) If $b^2 = 4ac$, the square root term is equal to zero, and the solution is simply

$$x = -\frac{b}{2a} \qquad \text{(1.27)}$$

(2) If $b^2 > 4ac$, we have two solutions, corresponding to the two possible signs of the square-root term. We call the solution with the positive sign x_+ and we use x_- to indicate the solution with the negative sign:

$$x_+ = \frac{-b + \sqrt{b^2 - 4ac}}{2a} \qquad \text{(1.28a)}$$

$$x_- = \frac{-b - \sqrt{b^2 - 4ac}}{2a} \qquad \text{(1.28b)}$$

(3) If $b^2 < 4ac$, we again have two solutions, but we must now take the square root of a *negative* quantity. Suppose that we have $\sqrt{-4}$. We can write this as

$\sqrt{(-1)(4)} = \sqrt{-1}\sqrt{4} = \sqrt{-1} \times (\pm 2)$. The quantity $\sqrt{-1}$ is called an *imaginary* quantity and is denoted by the symbol i:

$$\sqrt{-1} = i; \quad i^2 = -1 \tag{1.29}$$

Therefore, $\sqrt{-4} = \pm 2i$.

Imaginary quantities appear in the mathematical descriptions of certain types of physical problems, but we will not pursue such problems here and will not discuss imaginary quantities further.

▶ **EXAMPLE 1.13.1**   Solve the equation

$$x^2 + 4x + 4 = 0 \tag{1}$$

We identify

$$a = 1, \quad b = 4, \quad c = 4 \tag{2}$$

and we note that $b^2 = 4ac$. Then, using Equation (1.27), we have

$$x = -\frac{4}{2 \cdot 1} = -2 \tag{3}$$

The answer is readily verified to be correct by substitution of (3) into (1).

▶ **EXAMPLE 1.13.2**   Solve the equation

$$7x^2 - 8x + 1 = 0 \tag{1}$$

We identify

$$a = 7, \quad b = -8, \quad c = 1 \tag{2}$$

and we see that $b^2 = 64$ and $4ac = 28$, so that $b^2 > 4ac$. Using Equation (1.26), the solution is

$$x = \frac{-(-8) \pm \sqrt{(-8)^2 - 4 \cdot 7 \cdot 1}}{2 \times 7}$$

$$= \frac{8 \pm \sqrt{64 - 28}}{14} = \frac{8 \pm \sqrt{36}}{14} = \frac{8 \pm 6}{14} \tag{3}$$

The two solutions are

$$x_+ = \frac{8 + 6}{14} = 1 \tag{4}$$

$$x_- = \frac{8 - 6}{14} = \frac{2}{14} = \frac{1}{7} \tag{5}$$

Again, it is easy to verify by substitution into (1) that each solution is valid.

In the example above there were *two* possible solutions. In a physics problem, one can usually determine which solution is physically meaningful by examining the *physics* of the situation. (Sometimes *both* solutions are valid.) The following example demonstrates such a procedure.

▶ **EXAMPLE 1.13.3** The position x of a particle at a time t, moving in either direction along a straight line, is given by

$$x = x_0 + v_0 t + \frac{1}{2} a t^2 \tag{1}$$

where x_0 is the position at time $t = 0$ and v_0 is the velocity at $t = 0$, and where a is the acceleration of the particle.

Suppose that a ball is thrown vertically downward with a velocity of 60 ft/s from the top of a 100-ft building. If the ball is released at $t = 0$, when will it strike the ground?

If we measure x from the top of the building *downward*, then $x_0 = 0$, $x = 100$ ft, and $v_0 = 60$ ft/s. The acceleration experienced by the particle is g, the acceleration due to gravity, and has the value $g = 32$ ft/s^2 (see Equation 1.23).

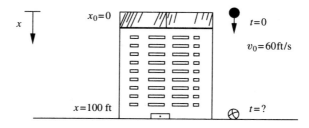

Writing (1) in standard form with $a = g$ and $x_0 = 0$, we have

$$\frac{1}{2} g t^2 + v_0 t - x = 0 \tag{2}$$

Solving for t,

$$t = \frac{-v_0 \pm \sqrt{v_0^2 - (4)\left(\frac{1}{2}g\right)(-x)}}{g} \tag{3}$$

Substituting the values for v_0, g, and x, we find

$$t = \frac{-60 \pm \sqrt{(60)^2 - (4)(16)(-100)}}{32}$$

$$= \frac{-60 \pm \sqrt{3600 + 6400}}{32}$$

$$= \frac{-60 \pm \sqrt{10^4}}{32}$$

$$= \frac{-60 \pm 100}{32} \tag{4}$$

The two possible solutions are

$$t_+ = \frac{-60 + 100}{32} = \frac{40}{32} = 1.25 \text{ s} \tag{5}$$

$$t_- = \frac{-60 - 100}{32} = -\frac{160}{32} = -5 \text{ s} \tag{6}$$

Clearly, the solution $t_- = -5$ s has no meaning in this case because the ball was released at $t = 0$. The proper solution is therefore $t_+ = 1.25$ s.

EXERCISES   *Solve the following equations:*

1. $3x^2 - 17x + 10 = 0$ (Ans. 359)
2. $x^2 - 2x - 24 = 0$ (Ans. 194)
3. $2x^2 - 5x - 7 = 0$ (Ans. 253)

4. $3x^2 - \dfrac{43}{2}x + \dfrac{7}{2} = 0$ (Ans. 179)
5. $x = x_0 + v_0 t + \frac{1}{2}gt^2$, where $x_0 = 0$, $x = 19.6$ m, $v_0 = 0$, and $g = 9.8$ m/s^2 (Ans. 268)

1.14   THE BINOMIAL EXPANSION

The multiplication of two algebraic quantities proceeds in the following way:

$$(a + b)(c + d) = a(c + d) + b(c + d)$$

$$= ac + ad + bc + bd$$

$$(a + b)^2 = (a + b)(a + b)$$
$$= a(a + b) + b(a + b)$$
$$= a^2 + ab + ba + b^2$$
$$= a^2 + 2ab + b^2$$

$$(a + b)^3 = (a + b)(a + b)^2$$
$$= (a + b)(a^2 + 2ab + b^2)$$
$$= a(a^2 + 2ab + b^2) + b(a^2 + 2ab + b^2)$$
$$= a^3 + 2a^2b + ab^2 + a^2b + 2ab^2 + b^3$$
$$= a^3 + 3a^2b + 3ab^2 + b^3$$

The general rule for the nth power of an algebraic quantity is given by the binomial theorem:

$$(a + b)^n = a^n + na^{n-1}b + \frac{n(n - 1)}{1 \cdot 2} a^{n-2}b^2$$
$$+ \frac{n(n - 1)(n - 2)}{1 \cdot 2 \cdot 3} a^{n-3}b^3 + \cdots \qquad \textbf{(1.30)}$$

Notice that when n is a positive integer this series does not continue indefinitely but terminates after $n + 1$ terms.

This expression permits writing the expansion of $(a + b)^3$ much more rapidly than does the method used above:

$$(a + b)^3 = a^3 + 3a^2b + \frac{3 \cdot 2}{1 \cdot 2} ab^2 + \frac{3 \cdot 2 \cdot 1}{1 \cdot 2 \cdot 3} a^0b^3$$
$$= a^3 + 3a^2b + 3ab^2 + b^3$$

We find many uses for the binomial expansion in physics problems. One of the most common cases is that in which $a = 1$. Then, Equation (1.30) becomes

$$(1 + b)^n = 1 + nb + \frac{n(n - 1)}{1 \cdot 2} b^2 + \frac{n(n - 1)(n - 2)}{1 \cdot 2 \cdot 3} b^3 + \cdots \qquad \textbf{(1.31)}$$

For example,

$$(1 + b)^3 = 1 + 3b + 3b^2 + b^3$$

We can always convert an expression of the type $(a + b)^n$ to one of the form $(1 + b)^n$ by *factoring* the original expression. For example,

$$(3 + x)^3 = 3^3\left(1 + \frac{x}{3}\right)^3 = 27(1 + y)^3$$

where we have used y to represent $x/3$.

Reciprocals can be obtained by using negative values of n:

$$(1 + b)^{-1} = 1 - b + b^2 - b^3 + \cdots \tag{1.32}$$

$$(1 + b)^{-2} = 1 - 2b + 3b^2 - 4b^3 + \cdots \tag{1.33}$$

In these cases, notice that the series of terms does *not* terminate as it does for positive integer values of n. That is, the expansion for $(1 + b)^{-1}$ or for $(1 + b)^{-2}$ consists of an *infinite series* of terms. Unless $|b| < 1$, each term in the series is larger than the preceding term and the series is of no value in evaluating the quantity. We say that the series *converges* only for $|b| < 1$.

▶ **EXAMPLE 1.14.1**   Obtain the series expansion for $(4 + x)^{-2}$.

First, we factor 4 out of the expression and substitute y for $x/4$.

$$(4 + x)^{-2} = (4)^{-2}\left(1 + \frac{x}{4}\right)^{-2} = \frac{1}{16}(1 + y)^{-2}$$

Then, using Equation (1.33)

$$\frac{1}{16}(1 + y)^{-2} = \frac{1}{16}(1 - 2y + 3y^2 - 4y^3 + \cdots)$$

If $y = 2$ (so that $x = 4y = 8$), we find

$$(4 + x)^{-2} = \frac{1}{16}(1 - 2 \cdot 2 + 3 \cdot 4 - 4 \cdot 8 + \cdots)$$

$$= \frac{1}{16}(1 - 4 + 12 - 32 + \cdots)$$

and the value of the series fluctuates widely as we add more and more terms. This series does not converge. On the other hand, if we have $y = 0.1$ (so that $x = 4y = 0.4$), then we find

$$(4 + x)^{-2} = \frac{1}{16}(1 - 2 \cdot 0.1 + 3 \cdot 0.01 - 4 \cdot 0.001 + \cdots)$$

$$= \frac{1}{16}(1 - 0.2 + 0.03 - 0.004 + \cdots)$$

$$\cong \frac{1}{16}(0.826) = 0.05162$$

Adding more terms will improve the result in the next decimal place, but the result will never differ greatly from 0.05162.

Expansions for the *roots* of quantities can be obtained by substituting the appropriate fractional value for n:

$$(1 + b)^{1/2} = 1 + \frac{1}{2}b - \frac{1}{8}b^2 + \frac{1}{16}b^3 - \cdots \qquad (1.34)$$

$$(1 + b)^{-1/2} = 1 - \frac{1}{2}b + \frac{3}{8}b^2 - \frac{5}{16}b^3 + \cdots \qquad (1.35)$$

(Verify these expansions.) Again, we must have $|b| < 1$ for convergence.

One of the important applications of Equations (1.34) and (1.35) is in the theory of relativity where we are frequently called upon to evaluate $\sqrt{1 - \beta^2}$ and $1/\sqrt{1 - \beta^2}$, where β is the ratio of the speed of a particle to the speed of light, $\beta = v/c$. Substituting $-\beta^2$ for b, we find

$$\sqrt{1 - \beta^2} = 1 - \frac{1}{2}\beta^2 - \frac{1}{8}\beta^4 - \frac{1}{16}\beta^6 - \cdots \qquad (1.36)$$

$$\frac{1}{\sqrt{1 - \beta^2}} = 1 + \frac{1}{2}\beta^2 + \frac{3}{8}\beta^4 + \frac{5}{16}\beta^6 + \cdots \qquad (1.37)$$

If the speed of the particle is much smaller than the speed of light, $v \ll c$, then $\beta^2 \ll 1$. For example, if $v = 3 \times 10^7$ m/s, we have (since $c = 3 \times 10^8$ m/s)

$$\beta = \frac{v}{c} = \frac{3 \times 10^7 \text{ m/s}}{3 \times 10^8 \text{ m/s}} = 10^{-1}$$

so that

$$\beta^2 = (10^{-1})^2 = 10^{-2}$$

$$\beta^4 = (10^{-1})^4 = 10^{-4}$$

$$\beta^6 = (10^{-1})^6 = 10^{-6}$$

Then, Equation (1.36) becomes

$$\sqrt{1 - \beta^2} = 1 - \frac{1}{2} \times 10^{-2} - \frac{1}{8} \times 10^{-4} - \frac{1}{16} \times 10^{-6} - \cdots$$

$$= 1 - 0.005 - 0.0000125 - 0.0000000625 - \cdots$$

$$= 0.9949874375 \cong 0.995$$

It is apparent that the terms after $-(\frac{1}{2})\beta^2$ do not influence the results to any significant extent. Therefore, if $\beta^2 \ll 1$, for all practical purposes we can use the approximate expressions,

$$\left.\sqrt{1 - \beta^2} \cong 1 - \frac{1}{2}\beta^2 \right\} \quad \quad \text{(1.38)}$$
$$\left.\frac{1}{\sqrt{1 - \beta^2}} \cong 1 + \frac{1}{2}\beta^2 \right\}\beta^2 \ll 1 \quad \quad \text{(1.39)}$$

▶ **EXAMPLE 1.14.2**   According to relativity theory, the mass of a particle depends on its speed according to the equation

$$m = \frac{m_0}{\sqrt{1 - \beta^2}} \quad \quad \text{(1)}$$

where m_0 is the mass of the particle at rest. What is the mass of a hydrogen atom which is moving with a speed of 10^8 m/s?

According to the result of Exercise 2 in Section 1.4, the mass of a hydrogen atom (at rest) is

$$m_0 = 1.67 \times 10^{-27} \text{ kg} \quad \quad \text{(2)}$$

We also have

$$\beta = \frac{v}{c} = \frac{10^8 \text{ m/s}}{3 \times 10^8 \text{ m/s}} = 3.3 \times 10^{-1} \quad \quad \text{(3)}$$

Using Equation (1.35), we can write

$$m \cong m_0 \left(1 + \frac{1}{2}\beta^2\right)$$

$$= (1.67 \times 10^{-27} \text{ kg}) \times \left[1 + \frac{1}{2}(3.3 \times 10^{-1})^2\right]$$

$$= (1.67 \times 10^{-27} \, \text{kg}) \times (1 + 0.05)$$

$$= 1.75 \times 10^{-27} \, \text{kg}$$

That is, the mass of the moving hydrogen atom is 1.05 times (or 5% greater than) the mass of a hydrogen atom at rest.

We have used the binomial expansion to enable us to make *approximate* calculations of certain quantities. There is nothing "wrong" in obtaining an approximate answer to a problem. In many cases, the approximate answer is quite sufficient for all practical purposes. For example, suppose that we wish to compute the time delay between the instant that an astronaut transmits a radio signal from the Moon and the instant when that signal is received by Mission Control on Earth. The time delay is

$$\text{Time delay} = \frac{\text{distance from Moon to Earth}}{\text{speed of radio signal}}$$

The speed of the radio signal is just the speed of any electromagnetic signal, namely, the speed of light. We know this speed to a precision of 100 m/s, so it requires *seven* significant figures to express this speed. Similarly, experiments in which laser signals from the Earth are reflected from mirrors on the Moon have increased our knowledge concerning the Earth-Moon distance to a comparable precision. Therefore, to compute the time delay would require the division of one seven-digit number by another seven-digit number, producing an answer that is precise to seven significant figures. To what practical use could we put such a number in this case? Absolutely none! So we might as well have computed the time delay using *approximate* values for the distance and the speed:

$$\text{Time delay} = \frac{3.84 \times 10^8 \, \text{m}}{3.00 \times 10^8 \, \text{m/s}} = 1.28 \, \text{s}$$

Surely this is as precise an answer as we would need for almost every purpose.

Even in pursuing a *precise* science, such as physics, there is nothing disreputable in making approximate calculations.

EXERCISES

1. Expand $(1 + x)^3$ (Ans. 276)

2. Expand $(1 + x)^{-3}$ (Ans. 305)

3. Examine Equation (1.34) and deduce the next term in the expansion.   (Ans. 76)

4. What is the approximate value of $(1.018)^{1/3}$?   (Ans. 356)

5. If $x \ll 1$, what is the approximate value of $\frac{1}{1 + x}$?   (Ans. 267)

6. If $x \gg 1$, what is the approximate value of $\frac{1}{1 + x}$?   (Ans. 215)

7. According to relativity theory, the length ℓ of an object in motion is related to the length ℓ_0 of the same object when at rest by $\ell = \ell_0 \sqrt{1 - \beta^2}$. If a meter stick moves with a velocity $v = 3 \times 10^7$ m/s, what will be its length?   (Ans. 153)

2

Analytical Geometry and Trigonometry

2.1   GRAPHS

One of the ways in which a visual impression is given of physical, economic, or other information is through the use of a *graph*. Many types of graphical presentations are possible. One of the simplest is the "pie" graph, shown in Figure 2.1. The entire "pie" represents 100 percent of some quantity, and the area of each "slice" indicates the fraction devoted to (or consumed by) a particular element of the whole. A pie graph does not give detailed (or *quantitative*) information unless the fractional size of each slice is given or is measured.

FIGURE 2.1   A "pie" graph showing the fraction of the total U.S. federal budget spent for various purposes in Fiscal Year 1991 (October 1, 1990 through September 30, 1991). The total federal budget "outlays" in Fiscal Year 1991 were approximately 1323 billion dollars, whereas the total federal "receipts" were about 1054 billion dollars, corresponding to a "deficit" of 269 billion dollars.

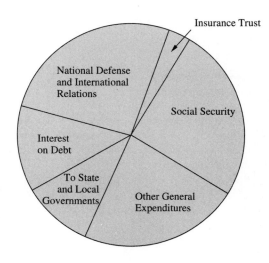

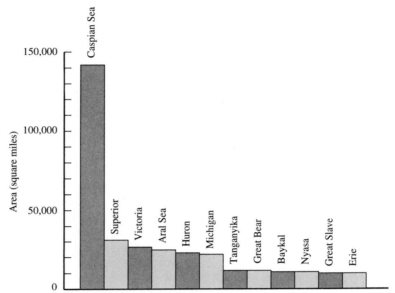

FIGURE 2.2 Comparison of the areas of the dozen largest lakes in the world.

Figure 2.2 shows a different kind of graph. Here, the vertical scale provides quantitative information—it represents the area of a lake in square miles. The horizontal direction, however, has no significance; each oblong box refers to one lake, in decreasing order of size. From this graph we can see that the Caspian Sea (which is actually a lake) is by far the largest inland body of water in the world with an area of approximately 144,000 sq mi. Lake Michigan, with an area of approximately 22,000 sq mi, is the largest lake that lies entirely within the United States.

The value of a quantity can be read only approximately from a graph such as that in Figure 2.2. If it were important to know that the area of Lake Victoria is 26,828 sq mi (instead of approximately 27,000 sq mi), then a list of geographical data should be consulted. However, certain types of information can be obtained with absolute accuracy from a graph.

Figure 2.3 shows the win-loss records of teams in the Eastern Division of the American Baseball League during the 1992 season. Again only the vertical scale carries quantitative information—the games won are plotted upward and the games

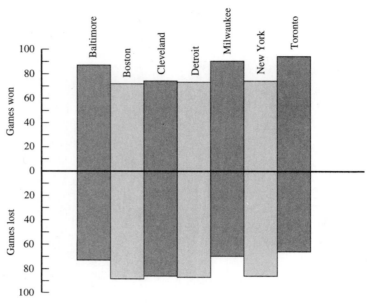

FIGURE 2.3   The 1992 win-loss records of teams in the Eastern Division of the American Baseball League. The Toronto Blue Jays went on to win the American League Championship Series, defeating Oakland 4 games to 2, and also defeated Atlanta in the World Series, 4 games to 2.

FIGURE 2.4   The world population from 1600 to the present.

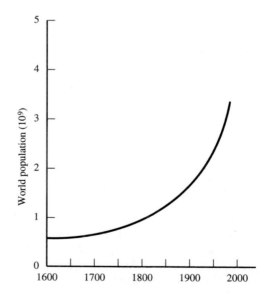

lost are plotted downward. The record of each club can be read precisely because there are no *fractional* games. For example, the Toronto Blue Jays won exactly 96 games and lost exactly 66 games during the 1992 season.

The next degree of complexity in a graph is to allow both the vertical and horizontal directions to carry quantitative information. Figure 2.4 shows the growth of the world population from about 1600 to the present time. The vertical scale indicates the population in billions of people (gigapeople), and the horizontal scale gives the time in years A.D. A graph such as this is imprecise for two reasons. First, the quantity of interest (the world population) is never known with high accuracy. Second, even if the population figure were known exactly, it could not be read from a graph that necessarily compresses a huge set of numbers into a small range.

EXERCISES

1. Refer to Figure 2.1. The fraction of the total area of the "pie" occupied by one of the "slices" can be obtained by measuring the angle between the two straight lines that define the slice and taking the ratio of this angle to 360° (which equals the complete circle). Use a protractor to measure the angle in the "social security" section and estimate the total expenditure for this item in Fiscal Year 1991.

(Ans. 164)

2. Estimate the ratio of the area of the Caspian Sea to that of the Aral Sea. (Use Fig. 2.2.)   (Ans. 382)
3. What club in the American League East had the poorest record in 1992? (Fig. 2.3.)   (Ans. 372)
4. What was the approximate world population at the time of the Second World War? (Fig. 2.4.)   (Ans. 7)

2.2   RECTANGULAR COORDINATES

Physical information is usually presented in a form similar to that used in Figure 2.4. The essential ingredients are a pair of *coordinate axes* and a *scale*. It is customary to use X for the horizontal axis (or *abscissa*) and Y for the vertical axis (or *ordinate*), and to allow both positive and negative values for each coordinate. Figure 2.5 illustrates the usual arrangement. X-Y systems of this type are usually called *rectangu-*

FIGURE 2.5   An *X-Y* rectangular coordinate system.

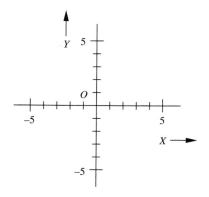

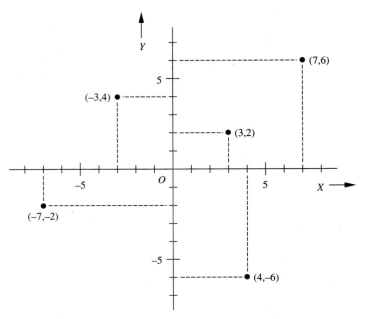

FIGURE 2.6   The location of points (x,y) in the X-Y plane.

lar coordinates, although the term *Cartesian coordinates* (after the French mathematician Descartes) is also used.

The location of a point in the X-Y coordinate system is specified by stating two numbers, the value of the X coordinate and the value of the Y coordinate; for example $(x = 3$ units, $y = 4$ units$)$. Usually, we simplify this procedure and write only the value of x and the value of y, it being understood that the order of the coordinates is first x and then y. Thus, the point referred to above is written as (3,4); a general point on the plane is written as (x,y). The origin is denoted by (0,0). Figure 2.6 locates several points in this notation.

Notice that we have not specified the magnitude of a unit of measure along the X and Y axes. Each unit could represent 1 cm, or 1 mile, or even 5.8 miles; in fact, the X unit could be different from the Y unit.

The X-Y coordinate system is useful for representing various types of physical situations, not only the location of points in terms of the distance from an origin. For example, suppose that we wish to represent graphically the length of a rubber band subject to various stretching forces. In this case, we would let the X-axis represent the *force* (and we would change X to F) and we would let the Y-axis represent the *length*. Figure 2.7 shows the measurements of the length, starting with $y = 0.10$ m, the unstretched length, for $F = 0$. Values are plotted at increments of 0.2×10^{-3} N of the stretching force and the curve is drawn through the points to give a smooth representation of the data. Notice that the length of the rubber band increases up to about 4 times its unstretched length, at which point it breaks.

We will pursue further the graphical representation of physical data in Chapter 3.

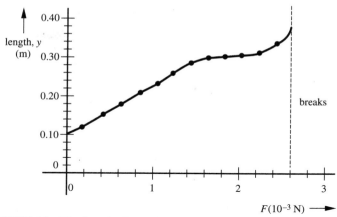

FIGURE 2.7 The length of a rubber band stretched by various forces.

EXERCISES *Plot the following sets of points on X-Y graphs and identify the geometrical shapes that the points outline:*

1. (0,0), (5,5), (0,5), (5,0) (Ans. 354) **3.** $(-2,1)$, $(6,9)$, $(-2,9)$, $(6,1)$ (Ans. 64)

2. (1,3), (3,9), (9,27), (27,51) (Ans. 137) **4.** $(0,-3)$, $(9,-3)$, $(9,0)$, $(0,0)$ (Ans. 349)

2.3 THREE-DIMENSIONAL COORDINATE SYSTEMS

The coordinate systems discussed in the preceding section are capable of representing the positions of points in a plane (i.e., in *two* dimensions, X and Y). If we wish to locate a point in *space,* such coordinate systems are no longer adequate because we require a third dimension. That is, if X and Y designate horizontal distances, we need also the *height* in order to locate a position in space. For this purpose we must add a third axis, Z, to our coordinate system, which now becomes a *three-dimensional* rectangular coordinate system, as shown in Figure 2.8.

Notice that the X-, Y-, and Z-axes in Figure 2.8 are oriented in a particular way. If one imagines that an ordinary (right-handed) screw is turned in the sense that carries the X-axis toward the Y-axis, the screw advances in the direction of the Z-axis (Figure 2.9). Such a system of coordinate axes is called a *right-handed* system and is the type that is used for almost all physical applications. (How would one define a *left-handed* coordinate system?)

The location of a point or the position of a physical object in space requires the specification of *three* numbers—the X, Y, and Z coordinates. If we wish to define a physical *event*, however, we need more information. An *event* is an occurrence that takes place not only at a particular location but also at a particular time. Therefore, the specification of an event requires three space coordinates *and* a time coordinate.

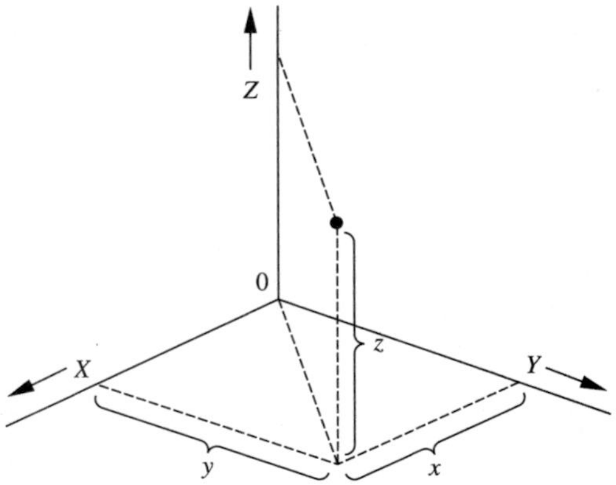

FIGURE 2.8   Location of the point (x,y,z) in a three-dimensional rectangular coordinate system.

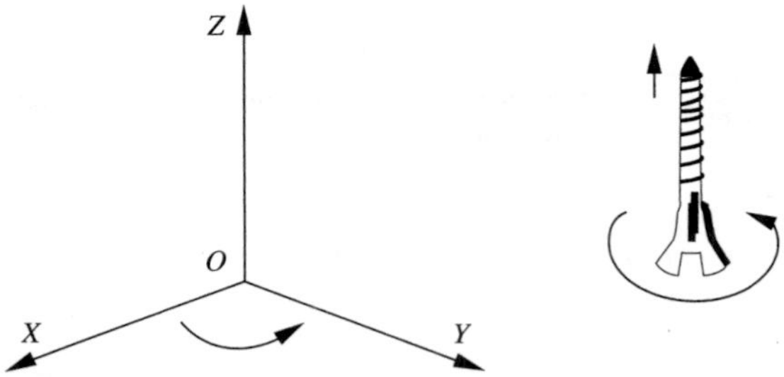

FIGURE 2.9   Definition of a *right-handed* coordinate system in terms of the direction of advance of a right-handed screw.

The location of an event (in *space* and in *time*) is given by *four* numbers and the coordinates are (x,y,z,t). That is, our space is actually a *four*-dimensional space with three ordinary space dimensions and one time dimension. Usually we do not think of our space as consisting of four dimensions because there is no apparent coupling or interrelation between ordinary space and time. But in the theory of relativity it becomes clear that there *is* such a coupling and that when objects are in motion with respect to one another with high relative velocities, the coupling of space and time is crucial. We should not lose sight of the fact that our world is, after all, a *four-dimensional* world.

EXERCISES

1. Construct a three-dimensional coordinate system and plot the following points: $(3,5,4)$, $(-2,4,-6)$, $(4,-3,7)$.

2. How would you plot an *event* in a four-dimensional coordinate system.   (Ans. 226)

3. Construct a space-time coordinate system for representing *events*. Restrict the space coordinates to *two* dimensions (X and Y). Plot two events that take place at the same location but at different times. Plot two events that take place at the same time but in different locations.

2.4   THE DISTANCE BETWEEN TWO POINTS

One of the important results of plane geometry is the *Pythagorean theorem,* which states that the square of the length of the side opposite the right angle (called the *hypotenuse*) of a right triangle is equal to the sum of the squares of the lengths of the other two sides. That is (refer to Figure 2.10),

$$c^2 = a^2 + b^2 \text{ (right triangle)} \tag{2.1}$$

Thus, if we know the lengths of any two sides of a right triangle, we can always find the length of the third side. If, in Figure 2.10, we know $c = 14$ cm and $a = 5$ cm, then

$$b = \sqrt{c^2 - a^2} = \sqrt{(14)^2 - (5)^2} \text{ cm} = \sqrt{196 - 25} \text{ cm}$$
$$= \sqrt{171} \text{ cm} = 13.08 \text{ cm}$$

FIGURE 2.10   Identification of sides in a right triangle.

▶ **EXAMPLE 2.4.1**   A motorist travels east from a starting point P at a constant speed of 50 km/h for 3 h. He then turns north and for 2 h he maintains a constant speed of 65 km/h. At the end of the 5-h trip, how far is he from P?

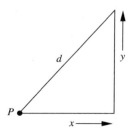

The first leg of the trip is

$$x = vt = (50 \text{ km/h}) \cdot (3 \text{ h}) = 150 \text{ km}$$

and the second leg is

$$y = (65 \text{ km/h}) \cdot (2 \text{ h}) = 130 \text{ km}$$

Therefore,

$$d = \sqrt{x^2 + y^2} = \sqrt{(150)^2 + (130)^2} \text{ km}$$
$$= \sqrt{22500 + 16900} \text{ km} = \sqrt{39400} \text{ km}$$
$$= \sqrt{3.94} \cdot 10^2 \text{ km} = 198.5 \text{ km}$$

Because the axes of a rectangular coordinate system are at *right* angles, the Pythagorean theorem can be used to calculate the distances between pairs of points in such a system. Consider the point A with coordinates (3,4) shown in Figure 2.11. How far is A from the origin? The point A and the origin define the right triangle OBA. Therefore, the distance c from O to A is

$$c = \sqrt{a^2 + b^2} = \sqrt{(3)^2 + (4)^2} = \sqrt{9 + 16}$$
$$= \sqrt{25} = 5 \text{ units}$$

Notice that this procedure works even if the coordinates of A are *negative* values because it is the *squares* of the coordinates that enter into the calculation.

If we wish to find the distance between two points neither of which is the origin,

FIGURE 2.11   The distance c from O to A can be found by using the Pythagorean theorem.

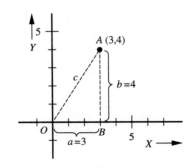

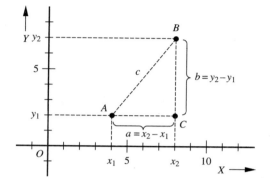

FIGURE 2.12   What is the distance from A to B?

then we follow a similar procedure. Such a case is shown in Figure 2.12 in which we require the distance from A to B. The two points A and B and the point C (which has the same x-value as B and the same y-value as A) define the right triangle ACB. Then, $c = \sqrt{a^2 + b^2}$, where

$$a = x_2 - x_1; \quad b = y_2 - y_1 \tag{2.2}$$

Therefore,

$$c = \sqrt{(x_2 - x_1)^2 + (y_2 - y_1)^2} \tag{2.3}$$

In Figure 2.12 the coordinates of the points are

$$A: (x_1, y_1) = (4,2); \quad B: (x_2, y_2) = (8,7)$$

Thus,

$$c = \sqrt{(8 - 4)^2 + (7 - 2)^2} = \sqrt{16 + 25}$$
$$= \sqrt{41} = 6.4 \text{ units}$$

In order to find the distance between two points in three-dimensional space, it is necessary to use the Pythagorean theorem *twice*. This procedure is best illustrated by means of an example.

▶ **EXAMPLE 2.4.2**   What is the length of the diagonal of a cube whose sides have a length of 1 m?
   The diagonal is the dashed line, OC. First, consider the right triangle OAB and the diagonal OB. Using the Pythagorean theorem, the length OB is

$$OB = \sqrt{(OA)^2 + (AB)^2} = \sqrt{(1)^2 + (1)^2} = \sqrt{2} \text{ m}$$

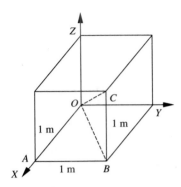

Next, consider the right triangle OBC. Again, we use the Pythagorean theorem to find the length of the side OC (the diagonal of the cube):

$$OC = \sqrt{(OB)^2 + (BC)^2} = \sqrt{(\sqrt{2})^2 + (1)^2} = \sqrt{3} \text{ m}$$

In general, the length of the diagonal of a cube is $\sqrt{3}$ times the length of a side.

The distance between two points located at (x_1, y_1, z_1) and (x_2, y_2, z_2) is found by combining the procedure of the preceding example and the result given in Equation (2.3); we find

$$\text{Distance} = \sqrt{(x_2 - x_1)^2 + (y_2 - y_1)^2 + (z_2 - z_1)^2} \qquad \text{(2.4)}$$

EXERCISES

1. What is the distance between the points (1,2) and (−6,3)?   (Ans. 42)

2. What is the length of the diagonal of a square whose area is 27 cm²?   (Ans. 160)

3. What is the distance from the origin to the point (1,2,3)?   (Ans. 58)

4. What is the length of the diagonal of a cube whose volume is 27 m³?   (Ans. 335)

2.5   RADIAN MEASURE

The most commonly used unit of angular measure is the *degree,* which is $\frac{1}{360}$ of a complete circle. For many types of problems in physics and engineering it proves

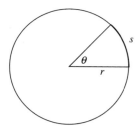

FIGURE 2.13 If $s = r$, the angle θ is equal to 1 radian.

more convenient to use another unit called the *radian* (rad). If we measure the length of arc along the circumference of a circle (see Figure 2.13), we find that the arc length s is proportional to the angle θ between the two radii that define the arc; that is, $s \propto \theta$. Furthermore, if we hold θ fixed and increase r, then s increases in direct proportion; that is, $s \propto r$. *One radian* is defined to be the angle subtended when the arc length s is exactly equal to the radius r. Thus,

$$s = r\theta \tag{2.5}$$

where θ is measured in radians.

If θ is increased until it is equal to $360°$, the arc s is just the circumference, $2\pi r$. Then, $s = 2\pi r = r\theta$, so that $\theta = 2\pi$ radians corresponds to $\theta = 360°$. Therefore,

$$1 \text{ rad} = \frac{360°}{2\pi} = 57.2958° \ldots \cong 57.3° \tag{2.6}$$

In order to find the radian equivalent of $1°$, we write

$$1° = \frac{2\pi}{360°} = 0.01745 \ldots \text{ rad} \tag{2.7}$$

Notice that although the *radian* is a unit of angular measure, it does not have physical dimensions. Therefore, when we put values into Equation (2.5), such as 3 cm = (3 cm) × (1 rad), we do not have different physical dimensions on the two sides of the equation. Usually, we carry the designation *radian* in our equations as a reminder of the angular units we are using, but when the final answer is obtained, it is sufficient to include only the *physical* dimensions. For example, if we calculate the length of the arc of a circle with $r = 10$ cm that is intercepted by an angle $\theta = 0.5$ radian, we obtain

$$s = r\theta = (10 \text{ cm}) \times (0.5 \text{ rad}) = 5 \text{ cm}$$

▶ **EXAMPLE 2.5.1** A 20-cm length of string is wrapped around a pipe and it is found that the ends of the string intercept an angle of 72° at the center of the pipe. What is the radius of the pipe?

First, the angle is

$$\theta = 72° = \frac{72°}{57.3°} \text{ rad} = 1.26 \text{ rad}$$

Then,

$$r = \frac{s}{\theta} = \frac{20 \text{ cm}}{1.26 \text{ rad}} = 15.9 \text{ cm}$$

The fact that $s = r\theta$ provides a method for closely estimating distances in certain circumstances. Suppose that a pole is placed vertically in the ground a distance R away from an observer located at a point O (Figure 2.14). The observer measures the angle subtended by the pole and finds it to be θ. Mentally, the observer constructs a circle of radius R centered at O and passing through the bottom of the pole. The arc length S is

$$S = R\theta \tag{2.8}$$

Since θ is a small angle, S is approximately equal to the height of the pole h, and we can write

$$h \cong R\theta \tag{2.9}$$

This method for obtaining approximate values for h will be useful if θ is sufficiently small. Even for θ as large as 20°, the error is only about 4 percent. If θ is a few degrees, the error is usually negligible for most purposes; for $\theta = 1°$, the error is 0.01 percent or 1 part in 10^4. (Remember that when using Equation 2.8 or 2.9 the angle θ must be expressed in *radians.*)

For a further discussion of this *small-angle approximation,* see Section 2.10.

FIGURE 2.14 Estimating the height of an object by measuring the angle subtended at O; $h \cong R\theta$.

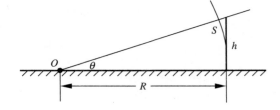

▶ **EXAMPLE 2.5.2** An observer on the Mall looks at the Washington Monument (height = 169 m) and determines that the Monument subtends an angle of 12°. How far away (approximately) is the observer from the Monument?

The angle is

$$\theta = 12° = \frac{12°}{57.3°} \text{ rad} = 0.21 \text{ rad}$$

Then,

$$R \cong \frac{h}{\theta} = \frac{169 \text{ m}}{0.21 \text{ rad}} = 8.06 \times 10^2 \text{ m}$$

or about 0.8 kilometer. The error in this approximate result is less than 1 percent.

EXERCISES

1. 320° = _____ rad   (Ans. 51)
2. 4 rad = _____ °   (Ans. 47)
3. 60° = _____ rad   (Ans. 320)
4. What angle does a 20-m tree subtend at a distance of 1 km?   (Ans. 330)

5. The diameter of the Moon is approximately 3.5×10^6 m, and the Moon subtends an angle of approximately $\frac{1}{2}°$ at the Earth. What is the Earth-Moon distance?   (Ans. 228)
6. The Empire State Building is 380 m high. An observer is 2 km from the building. What angle does the building subtend?   (Ans. 125)

2.6 ANGULAR MOTION

If an object moves in a circular path with a speed such that one complete revolution requires 1 second, we say that the object moves at an angular rate of 1 rev/s. Since one complete revolution corresponds to 2π radians, we can alternatively say that the object moves with an *angular* speed of 2π rad/s. It is customary to denote angular speed (measured in rad/s) by the symbol ω (omega).

If we do not have a complete revolution on which to base a calculation we can still define the angular speed in a manner entirely analogous to that used for ordinary (or *linear*) speed. Thus, if an object moves uniformly from a point identified by the

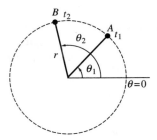

FIGURE 2.15   If the object moves uniformly from A to B, the angular speed is $(\theta_2 - \theta_1)/(t_2 - t_1)$ rad/s.

angle θ_1 to a point identified by θ_2 in a time interval $t_2 - t_1$ (see Figure 2.15), the *angular speed* is

$$\omega = \frac{\theta_2 - \theta_1}{t_2 - t_1} = \frac{\Delta\theta}{\Delta t} \qquad \text{(2.10)}$$

Since it is only the angular *difference* that is important, the position labeled $\theta = 0$ is arbitrary.

The *period* of circular motion is the time required for one complete revolution or cycle of the motion. The period and the angular speed are inversely related, since the greater the angular speed, the shorter the time required to make a revolution. The period is denoted by the symbol T:

$$T = \frac{2\pi}{\omega} \qquad \text{(2.11)}$$

If an object moves with uniform speed in a circular path with radius r, the distance traveled in 1 period is just the circumference of the circle, $2\pi r$. The time required for this motion is T. Therefore, the speed is

$$v = \frac{\text{distance}}{\text{time}} = \frac{2\pi r}{2\pi/\omega}$$

Thus,

$$v = r\omega \qquad \text{(2.12)}$$

▶ **EXAMPLE 2.6.1**   An automobile moves with a constant speed of 50 km/h around a circular track which has a diameter of 1 km. What is the angular speed and the period of the motion?

$$\omega = \frac{v}{r} = \frac{50 \text{ km/h}}{0.5 \text{ km}} = 100 \text{ rad/h}$$

$$T = \frac{2\pi}{\omega} = \frac{2\pi}{100 \text{ rad/h}} = 0.063 \text{ h} = 3.8 \text{ min}$$

EXERCISES

1. The Earth moves in a circular path about the Sun and completes 1 revolution in 365 days. What is the angular speed of the motion?   (Ans. 294)
2. The shaft of a certain electric motor rotates at a rate of 100 rpm. What is the angular speed in rad/s?   (Ans. 278)
3. In Exercise 2 above, what is the period of rotation of the shaft?   (Ans. 157)
4. An object moves in a circular path with a diameter of 1 m. The angular speed of the object is 2 rad/s. How far does the object move in one hour?   (Ans. 258)

5. What is the angular speed of the Earth's rotation around its axis?   (Ans. 53)
6. What is the speed of a point on the surface of the Earth at the equator due to the rotation of the Earth around its axis? (Use the fact that the radius of the Earth is approximately 6400 km and refer to the result of Exercise 5, above.) Express the result in km/s.   (Ans. 298)

2.7   TRIGONOMETRIC FUNCTIONS

A *right-triangle* is a triangle in which one of the angles is 90° (or $\pi/2$ radians). Such a right triangle is shown in Figure 2.16 in which the angle $\angle ACB$ is equal to 90°.

The lengths of the sides opposite the vertices *A, B,* and *C* are, respectively, *a, b,* and *c*. From the Pythagorean theorem (see Section 2.4), *a, b,* and *c* are related by

$$c^2 = a^2 + b^2 \tag{2.13}$$

We denote by θ the angle $\angle BAC$, and by ϕ the angle $\angle ABC$. The ratio of the length

FIGURE 2.16   Right triangle with $\angle ABC = 90°$.

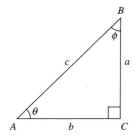

of the side opposite θ to the length of the hypotenuse (that is, side AB) is called the *sine* of the angle θ. This quantity is abbreviated as sin θ. Therefore,

$$\sin \theta = \frac{\text{side opposite angle } \theta}{\text{hypotenuse}} = \frac{a}{c} \tag{2.14}$$

Similarly, the ratio b/c is defined to be the *cosine* of θ; that is,

$$\cos \theta = \frac{\text{side adjacent angle } \theta}{\text{hypotenuse}} = \frac{b}{c} \tag{2.15}$$

Using Equation (2.13), we can also express sin θ and cos θ exclusively in terms of a and b:

$$\sin \theta = \frac{a}{\sqrt{a^2 + b^2}}; \quad \cos \theta = \frac{b}{\sqrt{a^2 + b^2}} \tag{2.16}$$

The ratio of the length of the side opposite θ to the length of the adjacent side is called the *tangent* of the angle θ:

$$\tan \theta = \frac{\text{side opposite angle } \theta}{\text{side adjacent angle } \theta} = \frac{a}{b} \tag{2.17}$$

The tangent of the angle θ is not independent of the sine and cosine since

$$\tan \theta = \frac{a}{b} = \frac{a/c}{b/c} = \frac{\sin \theta}{\cos \theta}$$

That is,

$$\tan \theta = \frac{\sin \theta}{\cos \theta} \tag{2.18}$$

In general, since b and a are each less than $c = \sqrt{a^2 + b^2}$, it follows from Equation (2.16) that sin θ and cos θ *do not exceed unity*. However, since tan θ = a/b, the tangent of the angle θ will exceed unity whenever a is larger than b.

▶ **EXAMPLE 2.7.1**   Consider a right triangle with a = 4 cm and b = 3 cm. In this case,

$$c = \sqrt{4^2 + 3^2} \text{ cm} = \sqrt{25} \text{ cm} = 5 \text{ cm}$$

and the triangle has the form shown in the accompanying figure. Therefore, from the defining equations for sin θ, cos θ, and tan θ, we find

$$\sin \theta = \frac{4 \text{ cm}}{5 \text{ cm}} = 0.800, \cos \theta = \frac{3 \text{ cm}}{5 \text{ cm}} = 0.600, \tan \theta = \frac{4 \text{ cm}}{3 \text{ cm}} = 1.333$$

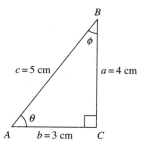

Note that the units of length cancel in the preceding expressions so that the resulting numbers for sin θ, cos θ, and tan θ are *dimensionless*. Note also that tan θ exceeds unity for this example.

The quantities sin θ, cos θ, and tan θ are called *trigonometric functions* of the angle θ. We sometimes see three additional trigonometric functions used, but these are all related directly to the original three:

$$\text{secant } \theta = \sec \theta = \frac{1}{\cos \theta}$$

$$\text{cosecant } \theta = \csc \theta = \frac{1}{\sin \theta}$$

$$\text{cotangent } \theta = \cot \theta = \frac{1}{\tan \theta}$$

EXERCISES   *If a = 1 m and b = 2 m, then*

1. c = _____   (Ans. 193)   **3.** cos θ = _____   (Ans. 324)
2. sin θ = _____   (Ans. 331)   **4.** tan θ = _____   (Ans. 166)

If c = 8 m and a = 2 m, then

5. b = _____   (Ans. 318)   **7.** cos θ = _____   (Ans. 108)
6. sin θ = _____   (Ans. 78)   **8.** tan θ = _____   (Ans. 203)

*(**Hint:** b can be determined by writing the Pythagorean theorem in the form b = $\sqrt{c^2 - a^2}$.)*

2.8   TRIGONOMETRIC IDENTITIES

We now summarize some of the important identities for trigonometric functions. First, we note that Equations (2.14) and (2.15) can be written as

$$c \sin \theta = a$$

$$c \cos \theta = b$$

If we square these equations and add them, we find*

$$\begin{array}{c} c^2 \sin^2 \theta = a^2 \\ c^2 \cos^2 \theta = b^2 \\ \hline c^2(\sin^2 \theta + \cos^2 \theta) = a^2 + b^2 \end{array}$$

Comparing this reult with Equation (2.13), we see that

$$\sin^2 \theta + \cos^2 \theta = 1 \qquad\qquad \textbf{(2.19)}$$

Therefore, if we know the value of sin θ, then cos θ can be determined from Equation (2.19), and *vice versa*. For example, if

$$\sin \theta = \frac{4}{5}$$

then it follows from Equation (2.19) that

$$\cos^2 \theta = 1 - \left(\frac{4}{5}\right)^2 = 1 - \frac{16}{25} = \frac{9}{25}$$

Taking the positive square root we find

$$\cos \theta = \frac{3}{5} = 0.6$$

which is in agreement with the result in Example 2.7.1.

There are some useful trigonometric relationships between the angles θ and ϕ in the right triangle shown in Figure 2.16. First, since the angles in a triangle add to 180°, we note that

$$\phi + \theta + 90° = 180°.$$

---

*It is customary to write $(\sin \theta)^2 = \sin^2 \theta$, and $(\cos \theta)^2 = \cos^2 \theta$. This is just a notational device, and no extra meaning should be read into it.

Therefore, the angles ϕ and θ are related by

$$\phi = 90° - \theta \tag{2.20}$$

The angle ϕ is called the *complement* of the angle θ. Similarly, since $\theta = 90° - \phi$, the angle θ is the complement of ϕ. The sine, cosine, and tangent of the angle ϕ are, by definition,

$$\sin \phi = \frac{b}{c}; \quad \cos \phi = \frac{a}{c}; \quad \tan \phi = \frac{b}{a}$$

Comparing these equations with the equations for $\sin \theta$, $\cos \theta$, and $\tan \theta$, we see that

$$\sin \theta = \frac{a}{c} = \cos \phi$$

$$\cos \theta = \frac{b}{c} = \sin \phi$$

$$\tan \theta = \frac{a}{b} = \frac{1}{\tan \phi} \tag{2.21}$$

Combining Equations (2.20) and (2.21) we find the identities,

$$\sin \theta = \cos(90° - \theta)$$

$$\cos \theta = \sin(90° - \theta)$$

$$\tan \theta = \frac{1}{\tan(90° - \theta)} \tag{2.22}$$

That is, *the sine of an angle is equal to the cosine of the complement of that angle; the cosine of an angle is equal to the sine of the complement of that angle;* and so on. Thus, for example,

$$\sin 20° = \cos 70°$$

$$\cos 50° = \sin 40°$$

$$\tan 80° = \frac{1}{\tan 10°}$$

$$\cos 63.5° = \sin 26.5°$$

Finally, we summarize without proof some additional relationships that are important in special situations. Sometimes it is necessary to find the trigonometric

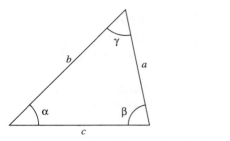
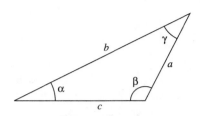

FIGURE 2.17 Skew triangles for which the relations in Equations (2.25) are valid.

functions for the sum or the difference of two angles. The following equations are then useful:

$$\sin(\alpha \pm \beta) = \sin \alpha \cos \beta \pm \cos \alpha \sin \beta$$

$$\cos(\alpha \pm \beta) = \cos \alpha \cos \beta \mp \sin \alpha \sin \beta$$

$$\tan(\alpha \pm \beta) = \frac{\tan \alpha \pm \tan \beta}{1 \mp \tan \alpha \tan \beta}$$

(2.23)

where the upper sign in $\pm$ or $\mp$ is carried through the equation if a function of the sum, $\alpha + \beta$, is desired, and where the lower sign is carried through for the difference, $\alpha - \beta$.

The *half-angle* formulas are also useful:

$$\sin \frac{1}{2}\alpha = \sqrt{\frac{1 - \cos \alpha}{2}}$$

$$\cos \frac{1}{2}\alpha = \sqrt{\frac{1 + \cos \alpha}{2}}$$

$$\tan \frac{1}{2}\alpha = \sqrt{\frac{1 - \cos \alpha}{1 + \cos \alpha}}$$

(2.24)

Although the trigonometric functions were defined in Equations (2.14), (2.15), and (2.17) in terms of the sides of a right triangle, these functions are associated with *angles* and not with particular triangles. Trigonometric functions can be used in a variety of situations, often involving triangles in which no angle is equal to 90°. Suppose that we have a skew triangle similar to one of those shown in Figure 2.17. The following relations are always valid:

$$\frac{\sin \alpha}{a} = \frac{\sin \beta}{b} = \frac{\sin \gamma}{c}$$

$$a^2 = b^2 + c^2 - 2bc \cos \alpha$$

(2.25)

1. If $\sin \theta = 1/4$, then $\cos^2 \theta =$ _____ (Ans. 260)

2. If $\sin \theta = 1/4$, then $\cos \theta =$ _____ (Ans. 171)

3. If $\cos \theta = 0.2$, then $\sin \theta =$ _____ (Ans. 43)

4. If $\cos \theta = 1/\sqrt{2}$, then $\tan \theta =$ _____

(Ans. 302)

5. The complement of $37°$ = _____ (Ans. 44)

6. The complement of $\pi/2$ radians = _____

(Ans. 283)

7. $\sin 23° = \cos$ _____ (Ans. 169)

8. $\cos 82° = \sin$ _____ (Ans. 95)

9. $\sin 65° = \cos$ _____ (Ans. 287)

10. $\tan 13° = \cot$ _____ (Ans. 168)

2.9 TABULATED TRIGONOMETRIC FUNCTIONS AND SOME IMPORTANT TRIANGLES

For easy reference, the values of the trigonometric functions $\sin \theta$, $\cos \theta$, and $\tan \theta$ are given in Table I in Appendix A for angles between $\theta = 0°$ and $\theta = 90°$, in steps of $1°$. Also given are the corresponding values of the angles in radians. Notice that the values of the functions are given to only three decimal places; this is sufficient for solving all examples and exercises in this book. However, students are urged to make use of an electronic calculator that provides much greater accuracy in the trigonometric functions at the push of a button. To obtain familiarity with the table and with your calculator (or both), verify the following:

For θ in degrees:

$$\sin 10° = 0.174 \qquad \sin 12° = 0.2079117$$

$$\cos 83° = 0.122 \qquad \cos 71° = 0.3255682$$

$$\tan 41° = 0.869 \qquad \tan 59° = 1.6642795$$

For θ in radians:

$$\sin 0.262 = 0.259 \qquad \sin 0.376 = 0.3672029$$

$$\cos 1.222 = 0.342 \qquad \cos 0.978 = 0.5586824$$

$$\tan 0.890 = 1.235 \qquad \tan 1.369 = 4.8880428$$

Also,

$$\sin \theta = 0.438 \text{ corresponds to } \theta = 26° \text{ or } 0.454 \text{ rad}$$

$$\cos \theta = 0.669 \text{ corresponds to } \theta = 48° \text{ or } 0.838 \text{ rad}$$

$$\tan \theta = 1.664 \text{ corresponds to } \theta = 59° \text{ or } 1.030 \text{ rad}$$

▶ **EXAMPLE 2.9.1**   A surveyor wishes to determine the distance between two points, A and B, but he cannot make a direct measurement because a river intervenes. How can he obtain a precise value for the distance?

The surveyor first selects a point C and he adjusts the location of C until a sighting with his transit (located at A) shows that the lines AB and AC are at right angles; that is, $\angle CAB = 90°$. He then measures the distance from A to C and finds

$$AC = 264 \text{ m}$$

Next, he positions his transit at C and sights toward A and then toward B, measuring the angle between the lines CA and CB. He finds $\angle ACB = 62.3°$
Now,

$$\tan \angle ACB = \frac{AB}{AC}$$

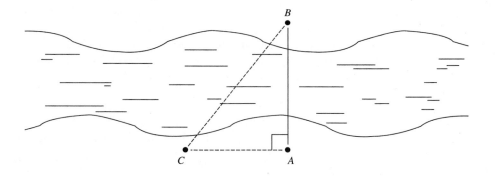

or, solving for the desired distance AB,

$$AB = AC \tan \angle ACB = (264 \text{ m}) \times (\tan 62.3°)$$

$$= (264 \text{ m}) \times (1.905) = 503 \text{ m}$$

Notice that the actual distance from A to C does not matter. That is, as long as $\angle CAB = 90°$ and the points A and B are fixed, the location of point C is irrelevant. The surveyor needs only to measure the distance AC and the angle $\angle ACB$; any combination will then produce the same result for AB.

Some triangles recur so frequently in physical problems that it is worthwhile to summarize the trigonometric functions for these cases. Figure 2.18 shows the 45°-45°-90° triangle for which the lengths of the sides are in the ratios $1:1:\sqrt{2}$;

FIGURE 2.18   The 45°-45°-90° triangle.

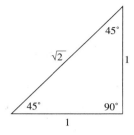

FIGURE 2.19   The 30°-60°-90° triangle.

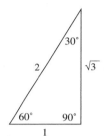

Figure 2.19 shows the 30°-60°-90° triangle for which the corresponding ratios are $1:\sqrt{3}:2$. It is easy to reproduce these ratios from memory (use the Pythagorean theorem) and so the trigonometric functions for the important angles of 30°, 45°, and 60° can be written down without the necessity of referring to Table I, Appendix A. The short table below summarizes the values of the trigonometric functions for all the angles involved in these simple triangles, plus the cases $\theta = 0°$ and $\theta = 90°$. The symbol ∞ in Table 2.1 means *indefinitely large* or *infinite*. Any non-zero number divided by 0 is infinite. Referring to Figure 2.16, we note that for $\theta = 90°$ we require $b = 0$. Therefore, $\tan 90° = a/0 = \infty$, as listed in Table 2.1.

TABLE 2.1   Values of Some Trigonometric Functions

	sin θ	cos θ	tan θ
$\theta = 0°$	0	1	0
$\theta = 30°$	0.5	$\dfrac{\sqrt{3}}{2} = 0.866$	$\dfrac{1}{\sqrt{3}} = 0.577$
$\theta = 45°$	$\dfrac{1}{\sqrt{2}} = 0.707$	$\dfrac{1}{\sqrt{2}} = 0.707$	1
$\theta = 60°$	$\dfrac{\sqrt{3}}{2} = 0.866$	0.5	$\dfrac{\sqrt{3}}{1} = 1.732$
$\theta = 90°$	1	0	∞

The operation of expressing an angle θ in terms of the value of its sine, cosine, or tangent is stated mathematically in the following manner. Referring to the right triangle in Figure 2.16, we can express θ as

$$\theta = \sin^{-1}\left(\frac{a}{c}\right) \tag{2.26}$$

This equation is equivalent to the statement:

$$\theta = \text{the angle whose sine is } \frac{a}{c} \tag{2.27}$$

The exponent -1 on the sine function in Equation (2.26) does *not* refer to the reciprocal; it is nothing more than a notational device that expresses the statement in Equation (2.27).

Since $\tan \theta = a/b$, and $\cos \theta = b/c$, we can also express θ as

$$\theta = \tan^{-1}\left(\frac{a}{b}\right) \tag{2.28}$$

or as

$$\theta = \cos^{-1}\left(\frac{b}{c}\right) \tag{2.29}$$

As an example, we note from Figure 2.19 (or Table 2.1) that

$$60° = \sin^{-1}\left(\frac{\sqrt{3}}{2}\right) = \cos^{-1}\left(\frac{1}{2}\right) = \tan^{-1}\left(\frac{\sqrt{3}}{1}\right)$$

or, equivalently,

$$60° = \sin^{-1}(0.866) = \cos^{-1}(0.5) = \tan^{-1}(1.732)$$

EXERCISES

1. $\cos 3° = $ _____   (Ans. 120)   **4.** $\sin 87° = $ _____   (Ans. 62)
2. $\sin 80° = $ _____   (Ans. 30)   **5.** $\cos 26.1° = $ _____   (Ans. 37)
3. $\tan 43° = $ _____   (Ans. 112)   **6.** $0.914 = \cos$ _____   (Ans. 133)

7. $2.145 = \tan$ _____   (Ans. 60)   **10.** $\cot^{-1}(1) =$ _____   (Ans. 246)
8. $0.357 = \cos$ _____   (Ans. 10)   **11.** $\sin^{-1}(0.577) =$ _____   (Ans. 11)
9. $0.075 = \sin$ _____   (Ans. 284)

2.10   SMALL ANGLE APPROXIMATIONS

In this section we derive approximate expressions for $\sin\theta$, $\cos\theta$, and $\tan\theta$, for situations in which the angle θ is much smaller than 90°. Consider the right triangle shown in Figure 2.20. We assume that side BC is much shorter in length than side AC; that is, $y \ll x$. In this case the angle θ is much smaller than 90°, and AC and AB are approximately equal in length, since

$$r = \sqrt{x^2 + y^2} \cong x$$

Referring to Figure 2.20, if we construct a circular segment with radius r and origin A (the dotted line in the figure), it is evident that BC is approximately equal to the arc length s; that is,

$$y \cong s$$

However, from the discussion in Section 2.5,

$$s = r\theta$$

where θ is expressed in radians. Therefore, we conclude that

$$y \cong r\theta$$

Combining the above results with the defining equations for $\sin\theta$, $\cos\theta$, and $\tan\theta$, we find, for small angles θ,

$$\sin\theta = \frac{y}{r} \cong \frac{r\theta}{r} = \theta$$

$$\cos\theta = \frac{x}{r} \cong \frac{r}{r} = 1$$

FIGURE 2.20   Right triangle with angle $\angle BAC$ much smaller than 90°.

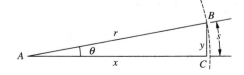

$$\tan \theta = \frac{y}{x} \cong \frac{r\theta}{r} = \theta$$

In summary, for small angles θ

$$\sin \theta \cong \theta$$
$$\cos \theta \cong 1$$
$$\tan \theta \cong \theta \qquad \qquad \text{(2.30)}$$

where θ is expressed in radians.

As an example, for $\theta = 2°$, we first convert to radians and find (to *three* significant figures)

$$\theta = \frac{2°}{360°} \times 2\pi = 0.0349 \text{ radians}$$

From Equations (2.30) this gives

$$\sin 2° \cong 0.0349$$
$$\cos 2° \cong 1$$
$$\tan 2° \cong 0.0349$$

These approximate values for sin 2°, cos 2°, and tan 2° agree with the values given in Table I.

When the angle θ is sufficiently large, the approximations in Equations (2.30) are no longer very accurate. For example, $\theta = 15°$ corresponds to

$$\theta = \frac{15°}{360°} \times 2\pi = 0.262 \text{ radians}$$

The approximations in Equations (2.30) then give

$$\sin 15° \cong 0.262$$
$$\cos 15° \cong 1$$
$$\tan 15° \cong 0.262$$

However, from the trigonometric tables in Table I, the actual values of sin 15°, cos 15°, and tan 15° are

$$\sin 15° = 0.259$$
$$\cos 15° = 0.966$$

$$\tan 15° = 0.268$$

Again, remember that approximate calculations are quite sufficient for many purposes. If your answer needs to be no more accurate than 3.5 percent, then the small-angle approximation for the sine, cosine, and tangent functions for angles less than 15° is entirely satisfactory. For all *precision* calculations, trigonometric tables should be consulted.

EXERCISES   *Using Equations **2.7** and **2.30**, evaluate*

1. sin 5° ≅ _____   (Ans. 132)   **4.** sin 0.1° ≅ _____   (Ans. 17)
2. tan 2.3° ≅ _____   (Ans. 296)   **5.** tan 3° ≅ _____   (Ans. 197)
3. cos 89° ≅ _____   (Ans. 316)

*Compare your answers with the corresponding values given in Table **2.1**.*

2.11   POLAR COORDINATES

The position of a point P in an X-Y coordinate system is specified by a pair of numbers (x,y), as shown in Figure 2.6. Alternatively, the equivalent position information can be given by specifying two other quantities, namely, the distance r from the origin O to the point P and the angle θ that the line OP makes with the X-axis (Figure 2.21). That is, we give the position of P in terms of the pair of numbers (r,θ) instead of the pair (x,y). The numbers r and θ are called the *polar coordinates* of the point.

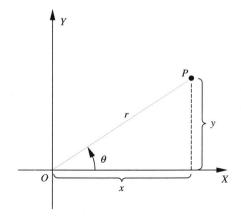

FIGURE 2.21   Polar coordinates (r,θ) of the point P which has coordinates (x,y) in a rectangular coordinate system.

As can be seen in Figure 2.21, the radial coordinate r of P is the hypotenuse of a right triangle in which x and y are the other two sides. Therefore, r (which is always *positive*) is given in terms of x and y by

$$r = +\sqrt{x^2 + y^2} \tag{2.31}$$

Using the definitions of the trigonometric functions, we can express the rectangular coordinates in terms of the polar coordinates:

$$x = r \cos \theta$$
$$y = r \sin \theta \tag{2.32}$$

Furthermore, using Equations (2.18) and (2.28), we can write

$$\tan \theta = \frac{y}{x}$$

$$\theta = \tan^{-1} \frac{y}{x} \tag{2.33}$$

▶ **EXAMPLE 2.11.1**   The polar coordinates of an object are ($r = 10$ m, $\theta = 60°$). What are the corresponding rectangular coordinates?

$$x = r \cos \theta = (10 \text{ m}) \cos 60° = (10 \text{ m}) \times 0.5 = 5 \text{ m}$$
$$y = r \sin \theta = (10 \text{ m}) \sin 60° = (10 \text{ m}) \times 0.866 = 8.66 \text{ m}$$

▶ **EXAMPLE 2.11.2**   A particle is located at the point ($x = 10$ cm, $y = 20$ cm) in a rectangular coordinate system. What are the corresponding polar coordinates?

$$r = \sqrt{(10)^2 + (20)^2} \text{ cm} = \sqrt{500} \text{ cm} = 22.36 \text{ cm}$$

We also have

$$\tan \theta = \frac{y}{x} = \frac{20 \text{ cm}}{10 \text{ cm}} = 2$$

Therefore,

$$\theta = \tan^{-1} 2 = 63.4°$$

This value can be estimated from Table I or can be found using an electronic calculator.

So far, we have restricted the discussion of polar coordinates to the quadrant of a rectangular coordinate system in which both x and y are positive quantities, and we have not discussed trigonometric functions for angles greater than 90°. It is clear, however, that there is no reason why a point cannot be located in other regions of a coordinate system. We usually specify these regions according to the labels shown in Figure 2.22. Quadrant II, for example, is the region from $\theta = 90°$ to $\theta = 180°$.

Equations (2.31), (2.32), and (2.33) are valid for the point P lying in any quadrant, but in all calculations we must be careful to incorporate the proper *signs* for x and y. (Recall that r is always *positive*.) For example, if P lies in Quadrant II, the coordinate y is positive, but the coordinate x is *negative* (Figure 2.23). Equations

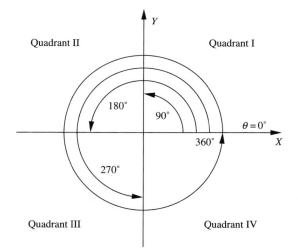

FIGURE 2.22   The quadrants in a coordinate system.

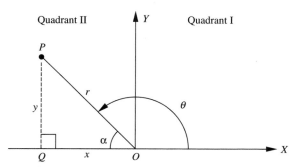

FIGURE 2.23   For P in Quadrant II, $y > 0$, but $x < 0$.

TABLE 2.2   Signs of the Coordinates and the Trigonometric Functions in the Various Quadrants

	Quadrant I	Quadrant II	Quadrant III	Quadrant IV
x	+	−	−	+
y	+	+	−	−
$\sin \theta$	+	+	−	−
$\cos \theta$	+	−	−	+
$\tan \theta$	+	−	+	−

(2.32) and (2.33) then tell us that the value of $\sin \theta$ is positive in Quadrant II but that the values of $\cos \theta$ and $\tan \theta$ are negative. Table 2.2 summarizes the results for all 4 quadrants.

In Figure 2.23, notice that the angle α (which is called the *supplement of* θ) is less than 90° *and that*

$$\alpha = 180° - \theta \qquad \text{(2.34)}$$

The sine of α is equal to y/r, and so is the sine of θ; that is,

$$\sin \theta = \sin \alpha = \sin(180° - \theta) \qquad \text{(2.35)}$$

To evaluate $\cos \theta$ requires some care. First, note that $\cos \alpha = |x|/r$. In the second quadrant, the value of x is negative, but α is less than 90° and so $\cos \alpha$ is a positive quantity. Therefore, in writing $|x|/r$ for $\cos \alpha$, we use the straight brackets that indicate the *magnitude* of x. Now, we also know that $\cos \theta = x/r$, which is a *negative* quantity. Therefore, we conclude that

$$\cos \theta = -\cos \alpha = -\cos(180° - \theta) \qquad \text{(2.36)}$$

Similarly, we find

$$\tan \theta = -\tan \alpha = -\tan(180° - \theta) \qquad \text{(2.37)}$$

Summarizing these results, for θ between 90° and 180° (Quadrant II):

$$\sin \theta = \sin(180° - \theta)$$
$$\cos \theta = -\cos(180° - \theta)$$
$$\tan \theta = -\tan(180° - \theta) \qquad \text{(2.38)}$$

Using similar reasoning, we can obtain the corresponding expressions for the remaining two quadrants:

For θ between 180° and 270° (Quadrant III):

$$\sin \theta = -\sin(\theta - 180°)$$
$$\cos \theta = -\cos(\theta - 180°)$$
$$\tan \theta = \tan(\theta - 180°) \qquad \text{(2.39)}$$

For θ between 270° and 360° (Quadrant IV):

$$\sin \theta = -\sin(360° - \theta)$$
$$\cos \theta = \cos(360° - \theta)$$
$$\tan \theta = -\tan(360° - \theta) \qquad \text{(2.40)}$$

Notice that in all of these equations, the trigonometric functions are expressed in terms of angles that are less than 90°. The values can therefore be found directly in the Trigonometric Tables and the signs are specified in the equations (and also in Table 2.2).

For example, using Equations (2.39), we find

$$\sin 250° = -\sin(250° - 180°) = -\sin 70° = -0.940$$

$$\cos 250° = -\cos(250° - 180°) = -\cos 70° = -0.342$$

$$\tan 250° = \tan(250° - 180°) = \tan 70° = 2.748$$

And, as a further example, for $\theta = 315°$, we use Equations (2.40) to find

$$\sin 315° = -\sin(360° - 315°) = -\sin 45° = -0.707$$

$$\cos 315° = \cos(360° - 315°) = \cos 45° = 0.707$$

$$\tan 315° = -\tan(360° - 315°) = -\tan 45° = -1.000$$

EXERCISES

1. For $y = 5$ m and $x = 2$ m, $r =$ _____
(Ans. 152)

2. For $y = 5$ m and $x = 2$ m, $\sin \theta =$ _____
(Ans. 329)

3. For $r = 2$ cm and $\theta = 45°$, $x =$ _____
(Ans. 23)

4. For $x = 4$ m and $r = 10$ m, $\theta =$ _____
(Ans. 192)

5. For $r = 10$ m and $x = 2$ m, $y =$ _____
(Ans. 45)

6. For $x = -5$ mm and $y = 15$ mm, $\tan \theta =$ _____
(Ans. 207)

7. For $y = 1$ km and $x = 100$ m, $r =$ _____

(Ans. 317)

8. $\sin 350° =$ _____   (Ans. 111)

9. $\cos 100° =$ _____   (Ans. 251)

10. $\tan 210° =$ _____   (Ans. 311)

11. $\tan 93.3° =$ _____   (Ans. 65)

12. $\cos 315° =$ _____   (Ans. 216)

13. $\sin 213° =$ _____   (Ans. 184)

14. $\cos 200.5° =$ _____   (Ans. 100)

2.12   GRAPHS OF TRIGONOMETRIC FUNCTIONS

It is instructive to examine the trigonometric functions in graphical form and to compare them with the results listed in Tables 2.1 and 2.2. Figure 2.24 shows the functions $\sin θ$, $\cos θ$, and $\tan θ$ for $θ$ from 0° to 360°. Notice that $\sin θ$ and $\cos θ$ vary between $+1$ and -1, whereas $\tan θ$ is unlimited. Notice further that $\sin θ$ and $\cos θ$ have exactly the same form. If we move the $\sin θ$ curve 90° to the left, we have exactly the $\cos θ$ curve. That is,

$$\cos (θ - 90°) = \sin θ$$

$$\sin (θ + 90°) = \cos θ \tag{2.41}$$

(Compare Equations 2.41 with Equations 2.22. Can both sets of equations be correct?)

The trigonometric functions are also defined for angles outside the range 0° to 360°. The angle $θ = -45°$, for example, is the same as the angle $θ = 315°$. If we start at $θ = 0°$ and describe an angle by performing two complete revolutions, we have an angle of 720° which is the same as 360° or 0°. That is, the trigonometric functions repeat every 360°, as shown in Figure 2.25 for $\sin θ$. The corresponding curve for $\cos θ$ can be obtained by shifting the $\sin θ$ curve by 90° in accordance with Equations (2.41).

In physics problems it is much more usual (and useful) to express $θ$ in radians instead of degrees. In Section 2.5 we showed that $360° = 2π$ rad, $180° = π$ rad, $90° = π/2$ rad, etc. Figure 2.26 illustrates the sine function again, with $θ$ given in radians.

Using radian measure, we can re-express Equations (2.41) as

$$\cos \left(θ - \frac{π}{2} \right) = \sin θ$$

$$\sin \left(θ + \frac{π}{2} \right) = \cos θ \tag{2.42}$$

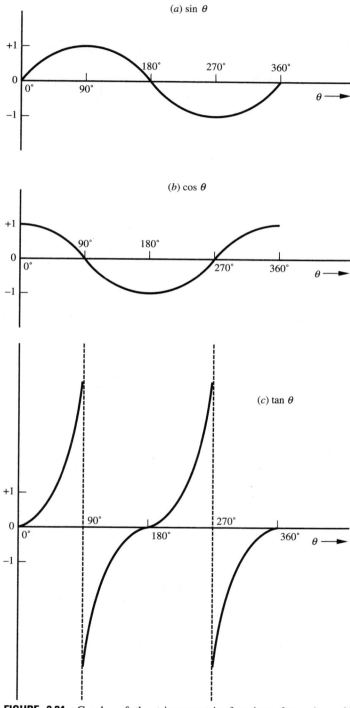

FIGURE 2.24 Graphs of the trigonometric functions from $\theta = 0°$ to $\theta = 360°$.

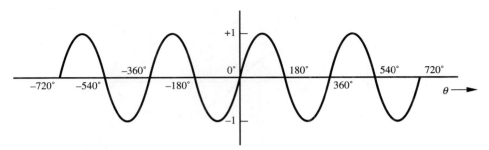

FIGURE 2.25   Sin θ for θ between $-720°$ and $+720°$.

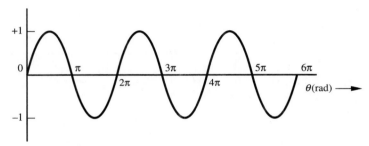

FIGURE 2.26   Sin θ for θ between 0 and 6π radians.

EXERCISES

1. Sin θ and cos θ vary between $+1$ and -1. Between what limits does tan θ vary?   (Ans. 368)

2. Evaluate sin θ and cos θ for $\theta = 135°$ using Equations (2.41) and Equations (2.38), and compare the results.

2.13   CYLINDRICAL AND SPHERICAL COORDINATES

In Section 2.3 we discussed the location of a point in space by using a three-dimensional rectangular coordinate system. For certain types of problems it is more convenient to specify a point in terms of radial and angular coordinates instead of in terms of *x, y,* and *z.* Therefore, we summarize here the results for *cylindrical* and *spherical* coordinates.

In a cylindrical coordinate system, the location of a point P amounts to constructing a cylinder whose axis is the Z-axis and whose radius is such that P lies

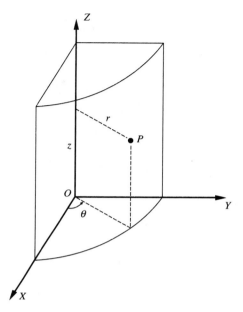

FIGURE 2.27 Cylindrical coordinates (r, θ, z).

on the surface of the cylinder (Figure 2.27). The three coordinates that locate P are (r, θ, z):

r = radius of the cylinder

θ = angle measured from the X-axis toward the Y-axis

z = height above the X-Y plane

The relationships connecting (r, θ, z) and (x, y, z) are:

$$x = r \cos \theta$$

$$y = r \sin \theta$$

$$z = z$$

$$r = \sqrt{x^2 + y^2}$$

$$\theta = \tan^{-1} \frac{y}{x} \tag{2.43}$$

Notice that cylindrical coordinates are exactly the same as polar coordinates except that the height coordinate z has been added.

In spherical coordinates, the location of a point P amounts to constructing a

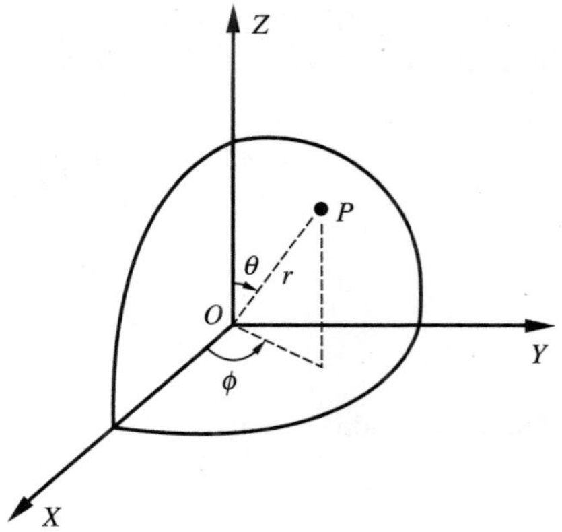

FIGURE 2.28 Spherical coordinates (r, θ, ϕ).

sphere whose center is the origin and whose radius is such that P lies on the surface of the sphere (Figure 2.28). The three spherical coordinates are:

r = radius of the sphere

θ = angle measured from the Z-axis toward the X-Y plane (the *polar* angle)

ϕ = angle measured from the X-axis toward the Y-axis (the *azimuthal* angle)

The relationships connecting (r, θ, ϕ) and (x, y, z) are:

$$x = r \sin \theta \cos \phi$$
$$y = r \sin \theta \sin \phi$$
$$z = r \cos \theta$$
$$r = \sqrt{x^2 + y^2 + z^2}$$
$$\theta = \cos^{-1} \frac{z}{r}$$
$$\phi = \tan^{-1} \frac{y}{x} \tag{2.44}$$

EXERCISES

1. What are the rectangular coordinates of a point whose spherical coordinates are (10, 30°, 60°)? (Ans. 170)

2. What are the cylindrical coordinates of the point whose rectangular coordinates are $(\sqrt{2}, \sqrt{2}, 5)$? (Ans. 214)

▶ 3

Functions and Graphs

3.1 CONCEPT OF A FUNCTION

In physics we are frequently called upon to find the *functional relationship* between different physical *variables*. For example, suppose that a particle moves along the X-axis as illustrated in Figure 3.1. We denote by x the particle's displacement from the origin O at the time t. As indicated in the figure, at time t_1 the particle's displacement is x_1. At some later time, t_2, the particle's displacement is x_2, and so on. That is, the particle's displacement (x) varies as a *function* of the time (t) that has elapsed. Mathematically, we describe this situation by writing

$$x = f(t) \tag{3.1}$$

Equation 3.1 simply states that "x is a function of t." The words "function of"

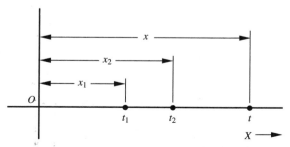

FIGURE 3.1 A particle's displacement (x) varies as a function of time (t).

are represented by the symbol $f(\ \ )$. For each value of t we can determine a corresponding value of x by measuring the particle's displacement from the origin. In Equation (3.1), the quantities x and t are referred to as *variables*. The variable t to which we first assign numerical values (t_1, t_2, etc.) is called the *independent variable*. The variable x, for which we determine the corresponding values, x_1, x_2, etc., is referred to as the *dependent variable*. That is, the displacement x *depends on* the time t that has elapsed.

We note from Figure 3.1 and Equation (3.1) that $x_1 = f(t_1)$, and $x_2 = f(t_2)$, and so on. The specific *form* of $f(t)$ in Equation (3.1) of course depends on the details of the particular problem. For example, we might have $x = at$, or $x = at + b$, or $x = \frac{1}{2} gt^2$, or even $x =$ constant. The right-hand sides of these equations represent different *functions* of the time t, and the expression that relates t to x is called a *functional relationship*. In the subsequent sections we will examine different types of functional relationships between a pair of variables.

3.2   REPRESENTATION OF FUNCTIONAL RELATIONSHIPS

The quantitative information collected in an experiment is referred to as *data*. These data are used to determine *functional relationships* between *variables* in the experiment. It is customary to represent these functional relationships by one or more of the following methods:

(1) By means of a *table*.
(2) By means of a *graph*.
(3) By means of an *equation*.

We now illustrate each of these methods with an example.

Suppose that the particle in Figure 3.1 moves in the positive X-direction with a *constant velocity* v_0 where $v_0 = 15$ m/s. Since the velocity is constant, this implies that during each 1-second time interval the particle traverses a distance of 15 m. As in Section 3.1 we represent the displacement of the particle from the origin O by the variable x. Furthermore, we denote the time elapsed by the variable t. For convenience, we take $x = 0$ at $t = 0$, by choosing the origin ($x = 0$) to coincide with the location of the particle at the time that we call $t = 0$.

The various methods of representing the functional relationship between the variables x and t for this example are as follows:

(1)   Representation by Means of a Table

Since the particle traverses 15 meters in each 1-second time interval, if we measure the displacement x of the particle we will find

- At the end of 1 second ($t = 1$ s), $x = 15$ m
- At the end of 2 seconds ($t = 2$ s), $x = 15$ m $+$ 15 m $= 30$ m
- At the end of 3 seconds ($t = 3$ s), $x = 15$ m $+$ 15 m $+$ 15 m $= 45$ m

TABLE 3.1 The Functional Relationship Between the Variables *x* and *t*

Time *t* (seconds)	Displacement *x* (meters)
0	0
1	15
2	30
3	45
4	60
5	75
6	90
7	105

and so on. In fact, at the end of N seconds ($t = N$ s), we will find $x = 15N$ meters. The values of t and the corresponding values of x are conveniently summarized (up to $t = 7$ s) in Table 3.1.

It is important to note that both the *symbols* (t and x) representing the variables and the *units* (seconds and meters, respectively) of these variables are clearly indicated in the table; this procedure should always be followed when tabulating data.

(2) Representation by Means of a Graph

Although we discussed coordinate systems and graphs in the preceding chapter, this subject is so important in physics that it is worthwhile ensuring that the techniques are well understood. In order to plot the data in Table 3.1, we choose a rectangular coordinate system with time measured along the horizontal axis and displacement measured along the vertical axis. Thus, we have a *t-x* graph. (Not every graph is an *x-y* graph!) Notice in Figure 3.2 that convenient scales have been included along the axes and that the units (*meters* for displacement and *seconds* for time) have been clearly indicated. This procedure should always be followed when plotting data.

To plot the data points which are tabulated in Table 3.1, we proceed as follows:

At $t = 0$ the value of x is $x = 0$. The coordinates of this point are (0,0), which is simply the *origin* in Figure 3.2.

According to Table 3.1, at $t = 1$ s the value of x is $x = 15$ m. The coordinates of this point are (1,15). As indicated by the first pair of dotted lines in Figure 3.2, the point (1,15) is located 1 s along the horizontal axis, and 15 m along the vertical axis.

From the table at $t = 2$ s the value of x is $x = 30$ m. The coordinates of this point are (2,30). As indicated by the second pair of dotted lines in Figure 3.2, the point (2,30) is located 2 s along the horizontal axis, and 30 m along the vertical axis.

In a similar manner, the data points (3,45), (4,60), (5,75), (6,90), and (7,105) are plotted in Figure 3.2.

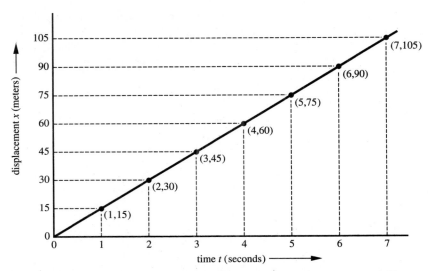

FIGURE 3.2   Graph representing the functional relationship between the variables x and t. Notice that each point has been labeled with coordinates as listed in Table 3.1.

In Figure 3.2 we have connected the data points by a solid line. Since a *single straight line* passes through *all* data points, the functional relationship between the variables x and t in the figure is known as a *linear relationship*. For this example the displacement x *increases linearly* with time t at a *constant* rate of 15 m/s.

(3)   Representation by Means of an Equation

We continue with the same example to illustrate the representation of the functional relationship by means of an equation. Since the displacement x increases with time t at a constant rate of 15 m/s, it is clear that we can relate the variables x and t by the equation

$$x = v_0 t \qquad \text{(with } v_0 = 15 \text{ m/s)} \tag{3.2}$$

In Equation (3.2) the units of x and t are *meters* and *seconds,* respectively. Furthermore, x is the *dependent variable,* t is the *independent variable,* and $v_0 = 15$ m/s is a *constant.* Equation (3.2) is the equation for a straight line and is a special example of the general functional relationship shown in Equation (3.1), for $f(t) = v_0 t$.

Notice that Equation (3.2) can be read in the following way:

> "The particle's displacement x (in meters) increases with time t (in seconds) at a constant rate of 15 m/s."

It is clear that Equation (3.2) can be used to reproduce both Table 3.1 and the graph in Figure 3.2. Since $v_0 = 15$ m/s, it follows from Equation (3.2) that

- At $t = 0$, $x = 0$
- At $t = 1$ s, $x = (15$ m/s$) \times (1$ s$) = 15$ m
- At $t = 2$ s, $x = (15$ m/s$) \times (2$ s$) = 30$ m
- At $t = 3$ s, $x = (15$ m/s$) \times (3$ s$) = 45$ m

and so on. These values of t and x are in agreement with the values given in Table 3.1 and Figure 3.2.

EXERCISES

1. From the following table construct a graph of x (in meters) versus t (in seconds).

Distance x (meters)	Time t (seconds)
0	0
2	0.5
4	1.0
6	1.5
8	2.0
10	2.5

2. What is the equation that relates the variables x and t in Exercise 1 above?   (Ans. 106)
3. What is the precise verbal statement that relates the variables x and t in Exercise 1 above?   (Ans. 289)

4. The displacement x (in kilometers) as a function of time t (in seconds) is given by the equation $x = v_0 t$, where $v_0 = 3$ km/s. Present the values of x and t from $t = 0$ to $t = 10$ s (in time intervals of 1 s) in tabular form.   (Ans. 86)
5. Construct a graph that relates the variables x and t in Exercise 4 above.
6. What is the precise verbal statement that relates the variables x and t in Exercise 4 above?   (Ans. 116)
7. Draw graphs of the following functions from $t = 0$ to $t = 10$ s:
   (a) $x = t$
   (b) $x = 2t$
   (c) $x = 3t$
   (d) $x = 4t$
   (e) $x = 5t$
   In each case x has units of meters, and t has units of seconds.

3.3   THE GENERAL STRAIGHT LINE

One of the simplest functional relationships that can exist between two variables occurs when one of the variables is directly proportional to the other. Such a functional relationship is said to be a *linear* relationship because it can be represented graphically by a *straight line*. The equation

$$x = v_0 t \tag{3.3}$$

considered in Section 3.2 is an example of a linear relationship.

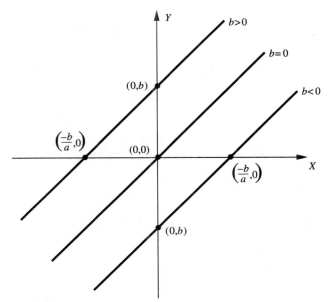

FIGURE 3.3   Graphs of $y = ax + b$ for $a > 0$ and for $b > 0$, $b = 0$, and $b < 0$.

In this section we consider the general case for a straight line represented by the equation

$$y = ax + b \qquad\qquad \text{(3.4)}$$

In contrast to Equation (3.3), the variable x in Equation (3.4) is the *independent variable*, and the variable y is the *dependent variable*. The quantities a and b in Equation (3.4) are *constants*. We do not specify the units of y, a, x, and b. However, since $y = ax + b$ is an *equation*, it should be noted that y, ax, and b have the *same* units.

Equation (3.4) is the most general type of *linear* relationship. In the particular case that $b = 0$, y is proportional to x.

The graphical representation of the straight line in Equation (3.4) is illustrated in Figures 3.3 and 3.4 for the case in which a is not equal to zero ($a \neq 0$). The independent variable x is plotted along the horizontal axis (the X-axis), and the dependent variable y is plotted along the vertical axis (the Y-axis). Now, from Equation (3.4) it follows that

- For $x = 0$, $y = a \cdot 0 + b = b$
- For $y = 0$, $0 = ax + b$, so that $x = -b/a$

That is, the straight line crosses the Y-axis (the line $x = 0$) at $y = b$, and the line

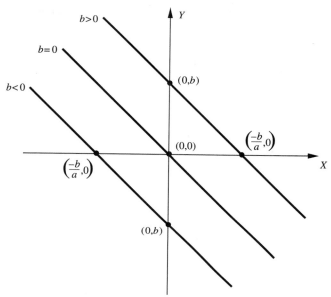

FIGURE 3.4 Graphs of $y = ax + b$ for $a < 0$ and for $b > 0$, $b = 0$, and $b < 0$.

crosses the X-axis ($y = 0$) at $x = -b/a$. These two points, $(0,b)$ and $(-b/a,0)$, are called the *intercepts* of the straight line; that is, the line intercepts the axes at these two points. Notice that a and b can be either *positive* or *negative* numbers.

The distinction between Figures 3.3 and 3.4 is that a is positive ($a > 0$) in Figure 3.3 and a is negative ($a < 0$) in Figure 3.4. Notice that lines are drawn for the three possible cases for b: $b > 0$, $b = 0$, and $b < 0$.

In Figure 3.3 (the top line) we have $a > 0$ and $b > 0$. Therefore, $-b/a$ is a *negative* quantity ($-b/a < 0$) and the straight line intercepts the X-axis to the *left* of the origin (i.e., at a *negative* value of x). Since $b > 0$, the line intercepts the Y-axis *above* the origin (i.e., at a *positive* value of y).

For each of the straight lines in Figures 3.3 and 3.4, note whether $-b/a$ is a positive or a negative quantity and similarly for b. Then, determine where the line will intercept each of the axes.

▶ **EXAMPLE 3.3.1** Draw a graph of the straight line $y = 2x + 4$.

For this example, $a = 2$ and $b = 4$. In addition, when $x = 0$, $y = 2 \cdot 0 + 4 = 4$. Therefore, the straight line $y = 2x + 4$ intersects the *positive* Y-axis at the point $(0,4)$. Furthermore, when $y = 0$, $2x + 4 = 0$; that is, $x = -4/2 = -2$. Therefore, the straight line $y = 2x + 4$ intersects the *negative* X-axis at the point $(-2,0)$. Thus, the straight line $y = 2x + 4$ has the form shown in the figure that follows and is similar to the general straight line represented in Figure 3.3.

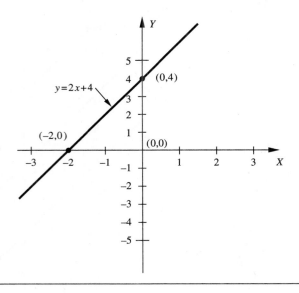

▶ **EXAMPLE 3.3.2**   Construct a graph of the equation $y = -2x$.

The equation $y = -2x$ is a special form of the general equation $y = ax + b$ with $b = 0$ and $a = -2$. In this case, if $x = 0$, then $y = 0$. That is, the straight line $y = -2x$ passes through the origin. In order to define the line, we need one additional point. If we choose $x = 1$, then the corresponding value of y is $y = (-2) \cdot 1 = -2$. Therefore, as shown in the figure below, the straight line $y = -2x$ passes through the point $(1, -2)$ as well as the origin $(0,0)$.

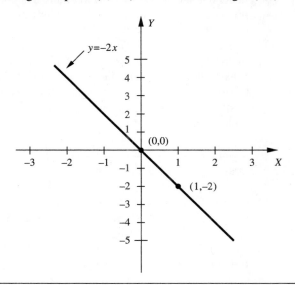

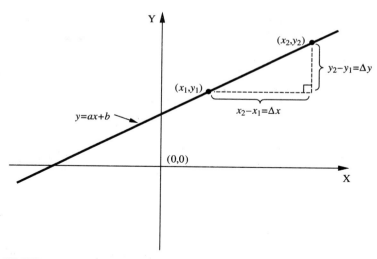

FIGURE 3.5 Graph of the straight line $y = ax + b$ that passes through the points (x_1, y_1) and (x_2, y_2).

An important characteristic of a straight line is its *slope* or inclination. Referring to Figure 3.5, we now consider a general straight line

$$y = ax + b$$

which passes through the two points (x_1, y_1) and (x_2, y_2). By definition, the *slope* of the straight line in Figure 3.5 is the rate at which y changes with x; that is,

$$\text{Slope} = \frac{y_2 - y_1}{x_2 - x_1} = \frac{\Delta y}{\Delta x} \tag{3.5}$$

The slope can also be expressed directly in terms of the constant a. Since the points (x_1, y_1) and (x_2, y_2) are located on the straight line $y = ax + b$, we conclude that

$$y_1 = ax_1 + b \tag{3.6}$$

and

$$y_2 = ax_2 + b \tag{3.7}$$

Subtracting Equation (3.6) from Equation (3.7), we find

$$y_2 - y_1 = a(x_2 - x_1)$$

Dividing through by $x_2 - x_1$ gives*

$$a = \frac{y_2 - y_1}{x_2 - x_1} \qquad \text{(3.8)}$$

Combining Equations (3.5) and (3.8), we can summarize the situation as follows:

> The slope of the straight line $y = ax + b$ is
>
> $$\text{Slope} = \frac{y_2 - y_1}{x_2 - x_1} = \frac{\Delta y}{\Delta x} = a \qquad \text{(3.9)}$$

Therefore, the slope of the straight line $y = 2x + 4$ (Example 3.3.1) is equal to 2, and the slope of the straight line $y = -2x$ (Example 3.3.2) is equal to -2.

In general, the straight line $y = ax + b$ has *positive* slope if $a > 0$. This is the case for each of the straight lines represented in Figure 3.3. Moreover, the straight line $y = ax + b$ has *negative* slope if $a < 0$. This is the case for each of the straight lines in Figure 3.4.

It is clear from Equation (3.5) that if the coordinates (x_1, y_1) and (x_2, y_2) are specified for *any* two points on a straight line, then the slope of the straight line can be calculated directly.

It is important to note the distinction between the "geometrical" and the "physical" interpretation of a slope. In the graph of Example 3.3.1, the straight line is at an angle of 45° with respect to the X-axis. Therefore, for every inch that we move along the X-direction, the line moves an inch in the Y-direction. Thus, the *geometrical* slope is *unity*. But notice that the physical units attached to the two scales are different. For every unit we move along the X-axis, the line moves upward by *two* units, and the *physical* slope is 2.

▶ **EXAMPLE 3.3.3**   What is the slope of a straight line passing through the points (1,2) and (11,8)?
Using the two points given, we have

$$x_1 = 1, \quad y_1 = 2$$

and

$$x_2 = 11, \quad y_2 = 8$$

Therefore, from Equation (3.5),

$$\text{Slope} = \frac{y_2 - y_1}{x_2 - x_1} = \frac{8 - 2}{11 - 1} = \frac{6}{10} = 0.6$$

---

*Of course, this is possible only if $x_2 - x_1 \neq 0$.

It is important to note that if we reverse the order of coordinate *labels,* then the value of the slope does not change. That is, if

$$x_1 = 11, \quad y_1 = 8$$

and

$$x_2 = 1, \quad y_2 = 2$$

then

$$\text{Slope} = \frac{y_2 - y_1}{x_2 - x_1} = \frac{2 - 8}{1 - 11} = \frac{-6}{-10} = \frac{6}{10} = 0.6$$

We now consider the special case in which a straight line has *zero slope.* In this case $a = 0$ and the equation $y = ax + b$ becomes

$$y = b \qquad\qquad \textbf{(3.10)}$$

That is, the value of y is equal to the constant b and is the same for *all values of x*. Equation (3.10) is an example of a *constant* functional relationship. The quantity y does not depend on the variable x and so y exhibits no change when the value of x changes. The straight line represented by Equation (3.10) is shown in Figure 3.6. Note that this line is parallel to the X-axis, and intersects the Y-axis at the point $(0,b)$. We reiterate that the straight line $y = b$ has *zero* slope, since $a = 0$ (see Equation 3.9).

Finally, we conclude this section by obtaining an expression for the constant b (Equation 3.4) in terms of the coordinates of any two points, (x_1,y_1) and (x_2,y_2), on

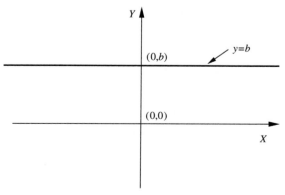

FIGURE 3.6 The straight line $y = b$. The slope is *zero.*

the general straight line $y = ax + b$. Eliminating a from Equation (3.6) by means of Equation (3.8), we find

$$y_1 = \left(\frac{y_2 - y_1}{x_2 - x_1}\right) x_1 + b$$

which gives

$$b = y_1 - \left(\frac{y_2 - y_1}{x_2 - x_1}\right) x_1 = \frac{y_1(x_2 - x_1) - (y_2 - y_1)x_1}{x_2 - x_1}$$

This expression for b further reduces to

$$b = \frac{y_1 x_2 - y_2 x_1}{x_2 - x_1} \tag{3.11}$$

▶ **EXAMPLE 3.3.4**   What is the equation of the straight line passing through the points (0,4) and (16,8)? We take

$$x_1 = 0, \quad y_1 = 4$$

and

$$x_2 = 16, \quad y_2 = 8$$

Therefore, from Equation (3.8),

$$a = \frac{y_2 - y_1}{x_2 - x_1} = \frac{8 - 4}{16 - 0} = \frac{4}{16} = \frac{1}{4} = 0.25$$

Furthermore, from Equation (3.11),

$$b = \frac{y_1 x_2 - y_2 x_1}{x_2 - x_1} = \frac{4 \cdot 16 - 8 \cdot 0}{16 - 0} = \frac{64}{16} = 4$$

Thus, the equation of the straight line passing through the points (0,4) and (16,8) is

$$y = 0.25x + 4$$

At the beginning of this section it was pointed out that the equation describing motion with constant velocity, $x = v_0 t$, is an example of a linear (or straight-line) relationship between the two variables, x and t. If a particle moves through a distance Δx in a time interval Δt, we can write

$$\Delta x = v_0 \Delta t \tag{3.12}$$

What is the interpretation of this equation? Dividing both sides of the equation by Δt, we have

$$v_0 = \frac{\Delta x}{\Delta t} \tag{3.13}$$

From Equation (3.5), we identify $\Delta x/\Delta t$ as the slope of the straight line defined by the equation, $x = v_0 t$. Therefore, we conclude that the slope of an x *versus* t graph for a particle is to be interpreted as the *velocity* of the particle. This is an important point, and we will return to the discussion of slopes and velocities in the following chapter.

EXERCISES

Construct graphs of the following straight lines:

1. $y = 2x + 3$

2. $y = -3x$

3. $y = 4x - 2$

4. $y = -x + 4$

What are the slopes of the following straight lines?

5. $y = -x + 4$ (Ans. 142)

6. $y = 7x$ (Ans. 107)

7. $y = 7x + 5$ (Ans. 326)

8. $y = -3x - 3$ (Ans. 55)

Determine the slopes of the straight lines which pass through the following points:

9. (2,5) and (36,5) (Ans. 128)

10. (1,3) and (3,1) (Ans. 201)

11. (16,17) and (17,18) (Ans. 139)

12. (0,0) and (67,68) (Ans. 148)

Determine the equations of the straight lines that pass through the following points:

13. (2,3) and (6,5) (Ans. 40)

14. (7,13) and (0,0) (Ans. 225)

15. (11,15) and (2,9) (Ans. 314)

16. (9,2) and (3,1) (Ans. 299)

3.4 UNIFORMLY ACCELERATED MOTION

In this section we discuss the equations that describe the motion of a particle of mass m along a straight line. We consider the particular case in which the particle

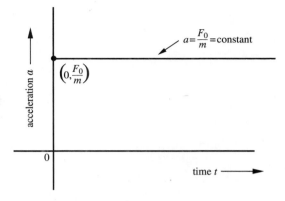

FIGURE 3.7 Graphical representation of acceleration a versus time t for uniformly accelerated motion (Equation 3.14).

experiences a *constant* force F_0 in the positive X-direction. According to Newton's Second Law, $F_0 = ma$; solving for a,

$$a = \frac{F_0}{m} = \text{constant} \tag{3.14}$$

The acceleration a is constant because the force is constant. Furthermore, the acceleration is in the same direction as the force, namely, the positive X-direction. Motion of this type is called *uniformly accelerated motion.*

We assume that at the time we designate $t = 0$ the particle is at the origin ($x = 0$) moving with a certain velocity v_0. As we will show in Chapter 6, using the methods of calculus, the velocity at any instant t is

$$v(t) = v_0 + at \tag{3.15}$$

and the distance x from the origin at time t is

$$x(t) = v_0 t + \frac{1}{2}at^2 \tag{3.16}$$

These are the functional relationships for uniformly accelerated motion. Notice that when calculations are made with these equations, a consistent set of units (e.g., SI units) must be used throughout.

We now examine the graphical representations of the three equations above that describe the motion. The graph of acceleration versus time (Figure 3.7) is an example of a *constant functional relationship*.

The velocity *versus* time graph (Figure 3.8) is a *straight line* with *slope equal to the acceleration a*. The straight line $v = v_0 + at$ intersects with the velocity axis at $(0, v_0)$. If $v_0 > 0$ (which corresponds to initial motion to the *right* in Fig. 3.1), then $v = v_0 + at$ intersects the *positive* velocity axis as shown in Figure 3.8.

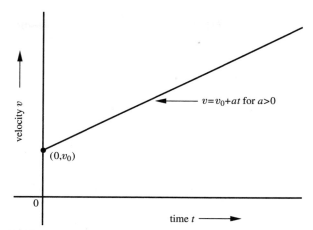

FIGURE 3.8   Graphical representation of velocity v versus time t for uniformly accelerated motion (Equation 3.15).

However, if $v_0 < 0$ (which corresponds to initial motion to the *left* in Fig. 3.1), then the straight line $v = v_0 + at$ intersects the *negative* velocity axis. In either case the slope of the straight line is *positive,* since $a > 0$.

The graph that gives the displacement x as a function of time (Equation 3.16) is shown in Figure 3.9 for the special case in which $v_0 = 0$. That is,

$$x(t) = \frac{1}{2}at^2 \quad (v_0 = 0) \tag{3.17}$$

FIGURE 3.9   Grapical representation of displacement x versus time t for uniformly accelerated motion with $v_0 = 0$ and $a > 0$ (Equation 3.17).

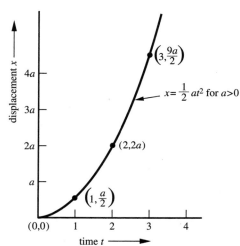

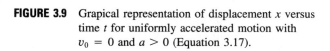

This is the equation for a *parabola*. Note that the units of x and t have not been specified in Figure 3.9 (nor in the previous figures). For $t = 1$, it is clear from Equation (3.17) that

$$x = \frac{1}{2} a \cdot (1)^2 = \frac{a}{2}$$

Hence, the point $(1, a/2)$ is located on the parabola in Figure 3.9. We leave it as an exercise for the reader to verify that $(2, 2a)$, $(3, 9a/2)$, $(-1, a/2)$, $(-2, 2a)$, and $(-3, 9a/2)$ are also located on the parabola whose equation is $x = \frac{1}{2} at^2$.

▶ **EXAMPLE 3.4.1**   A block of mass $m = 1$ kg is initially at rest ($v_0 = 0$) on a horizontal frictionless surface. A constant force $F_0 = 10$ N is applied to the block in the positive X-direction.

(a)   What is the acceleration of the block?
(b)   What is the velocity of the block after 10 s?
(c)   What is the displacement of the block after 10 s?

In answer to part (a) we find, from Equation (3.14),

$$\text{Acceleration} = a = \frac{F_0}{m} = \frac{10 \text{ kg} \cdot \dfrac{\text{m}}{\text{s}^2}}{1 \text{ kg}} = 10 \frac{\text{m}}{\text{s}^2}$$

In answer to part (b) we find, from Equation (3.15),

$$\text{Velocity} = v = v_0 + at = 0 + 10t = 10t \text{ (SI units)}$$

Therefore, for $t = 10$ s,

$$v = \left(10 \frac{\text{m}}{\text{s}^2}\right) \cdot (10 \text{ s}) = 100 \frac{\text{m}}{\text{s}}$$

In answer to part (c) we find, from Equation (3.16),

$$\text{Displacement} = x = v_0 t + \frac{1}{2} at^2 = 0 \cdot t + \frac{1}{2} \cdot 10t^2 = 5t^2 \text{ (SI units)}$$

Therefore, for $t = 10$ s,

$$x = \left(5 \frac{\text{m}}{\text{s}^2}\right) \cdot (10 \text{ s})^2 = 500 \text{ m}$$

▶ **EXAMPLE 3.4.2** As illustrated in the figure below, a man on the top of a building 49 m high leans over the edge of the building and drops a ball from rest ($v_0 = 0$). Determine

    (a)   the displacement of the ball at 1-second time intervals after the ball leaves the man's hand,

    (b)   the velocity of the ball at 1-second time intervals after the ball leaves the man's hand, and

    (c)   the time required for the ball to reach the pavement.

Take the acceleration due to gravity (g) to be exactly 9.8 m/s^2.

As indicated in the figure, we take the positive X-axis to be directed downward along the trajectory of the ball, with the origin O ($x = 0$) coinciding with the point at which the ball is released. Note that positive values of x and v are in the *downward* direction.

Since $a = g = 9.8$ m/s^2, and $v_0 = 0$, Equations (3.15) and (3.16) can be expressed as

$$\text{Velocity} = v = 0 + 9.8t = 9.8t \quad \text{(SI units)}$$

$$\text{Displacement} = x = 0 + \frac{9.8}{2}t^2 = 4.9t^2 \quad \text{(SI units)}$$

Therefore, in order to answer parts (a) and (b), we construct the following table:

Time t (seconds)	Displacement x (meters)	Velocity v (m/s)
0	0	0
1	4.9	9.8
2	19.6	19.6
3	44.1	29.4

In order to answer part (c), we solve (for t) the equation, $x = 4.9t^2$, when $x = 49$ m (the height of the building). This gives

$$49 = 4.9t^2$$

or,

$$10 = t^2$$

Therefore, the ball reaches the pavement in $t = \sqrt{10}$ s $= 3.162$ s.

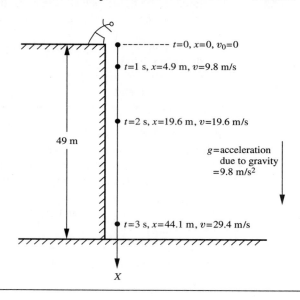

▶   **EXAMPLE 3.4.3**   As illustrated in the figure below, a woman on the top of a building 49 m high leans over the edge of the building and throws a ball vertically upward with an initial velocity $v_0 = -19.6$ m/s (the *negative* sign occurs since the initial motion is along the *negative* X-axis, as shown in the figure).

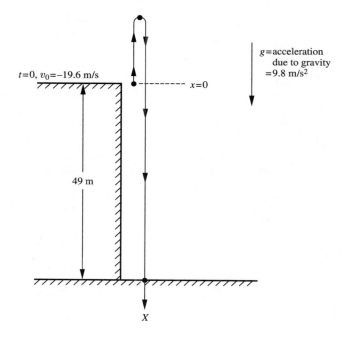

Determine

    (a)   the time required for the ball to reach maximum height above the top of the building,

    (b)   the maximum height to which the ball rises, and

    (c)   the time required for the ball to strike the pavement.

Take the acceleration due to gravity (g) to be exactly 9.8 m/s^2.

As in Example 3.4.2 we take the positive X-direction to be *downward,* with the origin O ($x = 0$) coinciding with the point at which the ball is released.

Since $a = g = 9.8$ m/s^2, and $v_0 = -19.6$ m/s, Equations (3.15) and (3.16) can be expressed as

$$\text{Velocity} = v = -19.6 + 9.8t \text{ (SI units)}$$

$$\text{Displacement} = x = -19.6t + 4.9t^2 \text{ (SI units)}$$

In answer to part (a), when the ball reaches the highest point on its trajectory, its velocity is equal to zero. This occurs at a time t determined from

$$v = 0 = -19.6 + 9.8t$$

Solving for t, we find that the ball reaches its maximum height at

$$t = \frac{19.6}{9.8} = 2 \text{ s}$$

In answer to part (b), to determine the maximum height to which the ball rises, we substitute $t = 2$ s into the equation $x = -19.6t + 4.9t^2$. This gives

$$x = -(19.6) \cdot 2 + 4.9 \cdot (2)^2$$
$$= -39.2 + 19.6$$
$$= -19.6 \text{ m}$$

Therefore, we conclude that the ball rises to a maximum height of 19.6 m above the top of the building. (It should be kept in mind that *negative* values of x correspond to displacements *above* the top of the building.)

In answer to part (c), we solve (for t) the equation, $x = -19.6t + 4.9t^2$, when $x = 49$ m (the *net* displacement of the ball when it strikes the pavement). This gives

$$49 = -19.6t + 4.9t^2$$

or

$$t^2 - 4t - 10 = 0$$

The solution to this quadratic equation for t is given by (see Equation 1.26)

$$t = \frac{4 \pm \sqrt{4^2 + 40}}{2}$$

We discard the solution with a negative value for t (just as in Example 1.13.3), and find

$$t = 2 + \sqrt{14} = 5.742 \text{ s}$$

EXERCISES

1. For Example 3.4.1, draw graphs of the acceleration, the velocity, and the displacement as functions of time from $t = 0$ to $t = 5$ s. (Compute the values at 1-s time intervals.)
2. Repeat, for Example 3.4.2.
3. Repeat, for Example 3.4.3.
4. What is the time required for the ball to pass by $x = 0$ in Example 3.4.3?   (Ans. 9)
5. What is the velocity of the ball as it passes by $x = 0$ in Example 3.4.3?   (Ans. 313)
6. A block of mass $m = 3$ kg moves on a horizontal frictional surface and experiences a constant force, $F_0 = +39$ N. The block has an initial velocity of $v_0 = -66$ m/s. How long does it take for the force F_0 to bring the block to rest?   (Ans. 181)
7. A man on the top of a building 230 m high leans over the edge of the building and releases a ball with downward velocity $v_0 = +10$ m/s. How long does it take for the ball to strike the pavement? (Take $g = 9.8$ m/s^2.)   (Ans. 119)
8. In Exercise 7 above, what is the velocity of the ball at the instant it strikes the pavement?   (Ans. 211)
9. In Exercise 7 above, what is the displacement of the ball 1 second after it is released?   (Ans. 66)
10. A car of mass $m = 1500$ kg experiences a constant forward thrust of $F_0 = +5000$ N. Assuming the car starts from rest, how long does it take for the car to reach a velocity of 20 m/s?   (Ans. 344)
11. In Exercise 10, what is the displacement of the car after 10 seconds?   (Ans. 158)
12. In Exercise 10, what is the velocity of the car after 10 seconds?   (Ans. 233)

3.5   SIMPLE HARMONIC MOTION

In this section we discuss the equations that describe the *simple harmonic* or *oscillatory* motion of a particle of mass m along a straight line. We consider the particular case in which the particle experiences a *restoring* force

$$F = -kx \tag{3.18}$$

which is *linearly* proportional to the displacement x from its equilibrium position at $x = 0$. Here, k is a positive constant, sometimes known as the "spring" constant. By analogy with the "back-and-forth" oscillatory motion of a mass m attached to a spring (see Figure 3.10), if the spring is extended a small distance x from its equilibrium position at $x = 0$, then the mass m experiences a restoring force $F = -kx$, which is linearly proportional to the displacement x. Here, k is the proportionality constant, and the *negative* sign in Equation (3.18) conveys that the force is restoring, that is *towards* $x = 0$. According to Newton's Second Law, $F = ma$; solving $ma = -kx$ for the acceleration a we find

$$a = -\frac{k}{m}x \tag{3.19}$$

It is customary to introduce the *angular frequency* of oscillation ω defined by

$$\omega = \sqrt{\frac{k}{m}} \tag{3.20}$$

so that Equation (3.19) can be expressed in the equivalent form

$$a = -\omega^2 x \tag{3.21}$$

Here, ω is related to the *period T* required for one complete oscillation, from $x = +x_0$ to $x = -x_0$ and back to $x = +x_0$ again in Figure 3.10, by the equation

$$T = \frac{2\pi}{\omega} \tag{3.22}$$

Unlike the case of uniformly accelerated motion considered in Section 3.4, we note from Equation (3.19) that the acceleration a is *not* constant [compare with Equation (3.14)]. Rather, for the case of simple harmonic motion, the acceleration a is linearly proportional to x, and therefore also depends on time t.

For simplicity, it is assumed in Figure 3.10 that the mass is released from rest (initial velocity $v_0 = 0$) with an initial *positive* displacement $x(t = 0) = x_0$. Then $F(t = 0) = -kx_0$ is *negative*, and the mass is first accelerated to the left, passes through $x = 0$ at time $t = T/4$, and then continues to a maximum negative displacement $x = -x_0$ at time $t = T/2$, where it experiences a positive force $F = -k(-x_0) = +kx_0$ and begins to move to the right. The mass again passes through $x = 0$ (at time $t = 3T/4$), and continues to a maximum positive displacement $x = +x_0$ at time $t = T$, the time period required for one complete oscillation. The motion then repeats, and the mass continues to oscillate between $+x_0$ and $-x_0$, and back to $+x_0$ again.

For the simple harmonic motion illustrated in Figure 3.10, the "initial" conditions at $t = 0$ are

$$x(t = 0) = x_0$$
$$v(t = 0) = v_0 = 0 \tag{3.23}$$

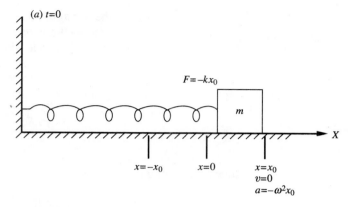

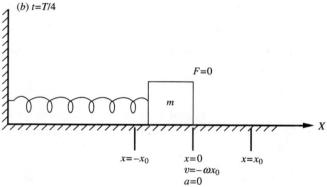

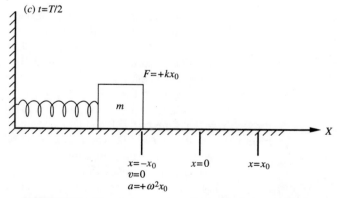

FIGURE 3.10 Simple harmonic motion of a mass m attached to a spring. The motion is shown for one-half period of a full oscillation from time $t = 0$ to $t = T/2$.

Using the methods of differential calculus (Chapter 6) it can be shown from Equation (3.19) that the displacement $x(t)$ at any instant of time t is

$$x(t) = x_0 \cos\left(\frac{2\pi t}{T}\right)$$

(3.24)

and the velocity $v(t)$ at time t is

$$v(t) = -\left(\frac{2\pi x_0}{T}\right) \sin\left(\frac{2\pi t}{T}\right)$$

$$= -\omega x_0 \sin\left(\frac{2\pi t}{T}\right)$$

(3.25)

where $\omega = \sqrt{k/m}$ is the angular frequency of oscillation, and $T = 2\pi/\omega$ is the period of one complete oscillation. From Equation (3.21), it follows that the acceleration is $a(t) = -\omega^2 x(t)$, which can also be expressed as

$$a(t) = -\left(\frac{2\pi}{T}\right)^2 x_0 \cos\left(\frac{2\pi t}{T}\right)$$

$$= -\omega^2 x_0 \cos\left(\frac{2\pi t}{T}\right)$$

(3.26)

where use has been made of $T = 2\pi/\omega$ and Equation (3.24).

Graphs of the trigonometric functions $\sin\theta$ and $\cos\theta$ plotted versus θ have been discussed in Section 2.12. Shown in Figure 3.11 are plots of the displacement $x(t)$, velocity $v(t)$, and acceleration $a(t) = -\omega^2 x(t)$ versus time t calculated from Equations (3.24), (3.25), and (3.26) for the "initial" conditions in Equation (3.23). Note from Figure 3.11 that the motion is *periodic* with periodicity time T. That is, the instantaneous values of displacement $x(t)$, velocity $v(t)$, and acceleration $a(t)$ recur at times $t = T$, $t = 2T$, $t = 3T$, etc. At time $t = 0$, the displacement x in Figure 3.11 begins at $x(t = 0) = x_0$ with velocity $v(t = 0) = 0$ and acceleration $a(t = 0) = -\omega^2 x_0$. As the mass is accelerated to the left, it passes through $x = 0$ at time $t = T/4$, where

$$\sin\left(2\pi \frac{t = T/4}{T}\right) = \sin\left(\frac{\pi}{2}\right)$$

$$= 1$$

and

$$\cos\left(2\pi \frac{t = T/4}{T}\right) = \cos\left(\frac{\pi}{2}\right)$$

$$= 0$$

where use is made of $\sin(\pi/2) = \sin(90°) = 1$ and $\cos(\pi/2) = \cos(90°) = 0$.

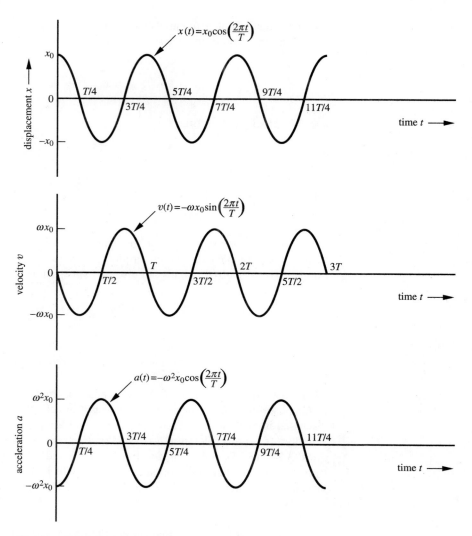

FIGURE 3.11. Graphical representation of the displacement x, velocity v, and acceleration a versus time t for the simple harmonic motion described by Equations (3.24), (3.25), and (3.26). The motion is plotted from time $t = 0$ to $t = 3T$.

From Equations (3.24), (3.25), and (3.26) this corresponds to

$$x(t = T/4) = 0$$

$$v(t = T/4) = -\omega x_0$$

$$a(t = T/4) = 0$$

At time $t = T/2$, we obtain

$$\sin\left(2\pi\frac{t = T/2}{T}\right) = \sin(\pi)$$

$$= 0$$

and

$$\cos\left(2\pi\frac{t = T/2}{T}\right) = \cos(\pi)$$

$$= -1$$

which from Equations (3.24), (3.25) and (3.26) give

$$x(t = T/2) = -x_0$$

$$v(t = T/2) = 0$$

$$a(t = T/2) = +\omega^2 x_0$$

The mass then accelerates to the right, and at time $t = 3T/4$ passes through $x(t = 3T/4) = 0$, with velocity $v(t = 3T/4) = +\omega x_0$ and acceleration $a(t = 3T/4) = 0$. Finally, from Equations (3.24), (3.25) and (3.26), at time $t = T$, the mass has returned to $x(t = T) = +x_0$, with velocity $v(t = T) = 0$ and acceleration $a(t = T) = -\omega^2 x_0$. The oscillatory motion then repeats, as illustrated in Figure 3.11.

▶ **EXAMPLE 3.5.1** A mass $m = 100$g is attached to a spring with spring constant $k = 9.869\text{g/s}^2 \approx \pi^2\text{g/s}^2$. The initial displacement of the mass is $x(t = 0) = x_0 = 5$ cm, and the initial velocity is $v(t = 0) = v_0 = 0$.

(a) What is the angular frequency of oscillation ω?
(b) What is the period T of one complete oscillation?
(c) What are the expressions for the displacement $x(t)$, velocity $v(t)$ and acceleration $a(t)$ that describe the oscillatory motion of the spring?
(d) What is the *maximum* speed of the mass as it passes through the origin ($x = 0$) at time $t = T/4$?

In answer to part (a), the angular frequency $\omega = \sqrt{k/m}$ calculated from Equation (3.20) is given by

$$\omega = \sqrt{\frac{\pi^2\text{g/s}^2}{100\text{g}}}$$

$$= \frac{\pi}{10}\text{ rad/s} \tag{1}$$

In answer to part (b), making use of the definition $T = 2\pi/\omega$ in Equation (3.22), we obtain for the period of oscillation

$$T = \frac{2\pi}{\pi/10}$$

$$= 20 \text{ s} \tag{2}$$

Because $x(t = 0) = x_0 = 5$ cm and $v(t = 0) = v_0 = 0$, it follows that $x(t)$, $v(t)$, and $a(t)$ are given by Equations (3.24), (3.25), and (3.26) with $x_0 = 5$ cm and $T = 20$s. We therefore obtain in answer to part (c)

$$x(t) = 5 \cos\left(\frac{\pi t}{10}\right) \tag{3}$$

$$v(t) = -\frac{\pi}{2} \sin\left(\frac{\pi t}{10}\right) \tag{4}$$

$$a(t) = -\frac{\pi^2}{20} \cos\left(\frac{\pi t}{10}\right) \tag{5}$$

where the units of x, v, and a are cm, cm/s and cm/s^2, respectively.

In answer to part (d), the maximum *speed* occurs as the mass passes through the origin $x = 0$ at times $t = T/4$, $t = 3T/4$, $t = 5T/4$, etc., where the period is $T = 20$s from the answer to part (b). Evaluating the magnitude of $v(t)$ for $t = T/4 = 5$s, we obtain

$$|v(t = 5\text{s})| = \left| -\frac{\pi}{2} \sin\left(\frac{\pi}{2}\right) \right|$$

$$= \left| -\frac{\pi}{2} \cdot 1 \right|$$

$$= \frac{\pi}{2} \text{ cm/s}$$

corresponding to a maximum speed of

$$|v(t = 5\text{s})| = 1.57 \text{ cm/s} \tag{6}$$

The "initial" conditions at time $t = 0$ in Equation (3.23) correspond to extending the spring with the mass m attached by a distance x_0 from the origin and releasing it from rest with $v(t = 0) = v_0 = 0$. A more general set of initial conditions corresponds to

$$x(t = 0) = x_0$$
$$v(t = 0) = v_0 \tag{3.27}$$

where x_0 and v_0 are both allowed to be non-zero. In this case, making use of the linear force relation $a = -(k/m)x = -\omega^2 x$ in Equations (3.19) and (3.20), the methods of differential calculus (Chapter 6) show that the displacement $x(t)$ and velocity $v(t)$ can be expressed as

$$x(t) = A \cos\left(\frac{2\pi t}{T} + \phi\right) \tag{3.28}$$

and

$$v(t) = -\frac{2\pi A}{T} \sin\left(\frac{2\pi t}{T} + \phi\right)$$
$$= -\omega A \sin\left(\frac{2\pi t}{T} + \phi\right) \tag{3.29}$$

where ω is the angular frequency, $T = 2\pi/\omega$ is the period of oscillation, and A and ϕ are constant amplitude and phase factors, respectively. Making use of $a(t) = -\omega^2 x(t)$, we obtain from Equations (3.21) and (3.28)

$$a(t) = -\left(\frac{2\pi}{T}\right)^2 A \cos\left(\frac{2\pi t}{T} + \phi\right)$$
$$= -\omega^2 A \cos\left(\frac{2\pi t}{T} + \phi\right) \tag{3.30}$$

for the acceleration $a(t)$.

As before, the motion described by Equations (3.28), (3.29) and (3.30) is *simple harmonic* with oscillation period $T = 2\pi/\omega$. Furthermore, the amplitude constant A and the phase constant ϕ can be related to the initial displacement x_0 and the initial velocity v_0 by means of Equation (3.27). Setting $t = 0$ in Equations (3.28) and (3.29), we obtain

$$x_0 = A \cos \phi$$
$$v_0 = -\omega A \sin \phi \tag{3.31}$$

From Equation (3.31), we note that the special case where the phase angle is $\phi = 0°$ corresponds to an initial velocity $v_0 = -\omega A \sin 0° = 0$, and an initial displacement $x_0 = A \cos 0° = A$. In this case, Equations (3.28)–(3.30) reduce exactly to Equations (3.24)–(3.26) with $\phi = 0$ and $x_0 = A$.

In the general case, Equation (3.31) can be used to express ϕ and A directly in terms of x_0 and v_0. Using $x_0 = A \cos \phi$ and $v_0/\omega = -A \sin \phi$, and taking the square root of the sum of the squares of both sides of the equations, we obtain

$$(x_0{}^2 + v_0{}^2/\omega^2)^{1/2} = (A^2 \cos^2\phi + A^2 \sin^2\phi)^{1/2}$$
$$= A(\cos^2\phi + \sin^2\phi)^{1/2}$$

Because $\cos^2\phi + \sin^2\phi = 1$ [see Equation (2.19)], this reduces to

$$A = (x_0{}^2 + v_0{}^2/\omega^2)^{1/2} \tag{3.32}$$

which relates the amplitude factor A directly to x_0 and v_0. Similarly, taking the ratio of both sides of the equations $A \sin \phi = -v_0/\omega$ and $A \cos \phi = x_0$ gives

$$\frac{A \sin \phi}{A \cos \phi} = \frac{-v_0/\omega}{x_0}$$

or equivalently

$$\tan \phi = -\frac{v_0}{\omega x_0} \tag{3.33}$$

where $\tan \phi = (\sin \phi)/(\cos \phi)$.

A detailed comparison of the equations for $x(t)$, $v(t)$, and $a(t)$ given in Equations (3.28)–(3.30) with those given in Equations (3.24)–(3.26) shows that the main difference in the simple harmonic motion described by the two sets of equations is a *shift in time origin* that depends on the phase angle ϕ. In both cases, the motion is periodic with time period T and oscillation frequency $\omega = 2\pi/T$. In both cases, the motion is simple harmonic with constant oscillation amplitude.

We illustrate this point with an example.

▶ **EXAMPLE 3.5.2**   An object with mass m is attached to a spring with spring constant k and undergoes simple harmonic motion along a straight line according to Equations (3.28), (3.29) and (3.30). At time $t = 0$, the mass is passing through the origin with initial position and velocity given by

$$x(t = 0) = x_0 = 0 \tag{1}$$

$$v(t = 0) = v_0 \tag{2}$$

where v_0 is non-zero and *negative* ($v_0 < 0$). That is, the mass is moving to the *left* at time $t = 0$.

(a) Derive expressions for the oscillation amplitude A and ϕ in terms of the initial velocity (v_0) and the initial displacement ($x_0 = 0$).

(b) What are the resulting expressions for $x(t)$, $v(t)$ and $a(t)$?

(c) Present plots of $x(t)$, $v(t)$ and $a(t)$ versus time t where the units of time are measured in terms of the oscillation period $T = 2\pi/\omega$.

In answer to part (a), we set $x_0 = 0$ in Equation (3.31). This gives

$$0 = A \cos \phi \tag{3}$$

$$v_0 = -\omega A \sin \phi \tag{4}$$

which will determine the amplitude A and phase angle ϕ. Because v_0 is non-zero, we must generally choose a non-zero oscillation amplitude A. Therefore, $A \cos \phi = 0$ gives

$$\cos \phi = 0 \tag{5}$$

and we conclude that $\phi = \pi/2$, or $\phi = 3\pi/2$, or $\phi = 5\pi/2$, or $\phi = 7\pi/2$, etc. (see Chapter 2). Because $v_0 < 0$ is assumed (the initial motion is to the *left*), and because $\sin(3\pi/2) = -1$, $\sin(7\pi/2) = -1$, etc., we conclude that only the phase angles

$$\phi = \frac{\pi}{2}$$

$$\phi = \frac{5\pi}{2} = \frac{\pi}{2} + 2\pi$$

$$\phi = \frac{9\pi}{2} = \frac{\pi}{2} + 4\pi$$

are allowed as solutions when $x_0 = 0$ and $v_0 < 0$. Furthermore, because the sine and cosine functions are *periodic* with period equal to 2π, we may choose the phase angle ϕ to be

$$\phi = \frac{\pi}{2}$$

without loss of generality. This follows because the sine and cosine functions exactly reproduce the same values when the argument is shifted by 2π, 4π, 6π, etc.

Setting $\phi = \pi/2$ in the expression $v_0 = -\omega A \sin \phi$ gives

$$v_0 = -\omega A \sin(\pi/2)$$

$$= -\omega A$$

where use is made of $\sin(\pi/2) = \sin(90°) = 1$. Solving $v_0 = -\omega A$ for the amplitude A gives the desired result

$$A = -\frac{v_0}{\omega} \tag{7}$$

In answer to part (b), we substitute $\phi = \pi/2$ and $A = -v_0/\omega$ into Equations (3.28), (3.29) and (3.30). This gives for the displacement $x(t)$, velocity $v(t)$ and acceleration $a(t)$

$$x(t) = -\frac{v_0}{\omega} \cos\left(\frac{2\pi t}{T} + \frac{\pi}{2}\right) \tag{8}$$

$$v(t) = v_0 \sin\left(\frac{2\pi t}{T} + \frac{\pi}{2}\right) \tag{9}$$

$$a(t) = \omega v_0 \cos\left(\frac{2\pi t}{T} + \frac{\pi}{2}\right) \tag{10}$$

where $\omega = 2\pi/T$ is the angular frequency of oscillation. Note from the expressions for $a(t)$ and $x(t)$ that $a(t) = -\omega^2 x(t)$, as expected.

In answer to part (c), shown in Figure 3.5.2 are plots of the displacement $x(t)$, velocity $v(t)$ and acceleration $a(t)$ versus time t. Graphs of the trigonometric functions $\sin \phi$ and $\cos \phi$ versus ϕ have been discussed in Section 2.12.

Note from Figure 3.5.2 that the motion is periodic with periodicity time T. That is, the instantaneous values of $x(t)$, $v(t)$ and $a(t)$ all recur at times $t = T$, $t = 2T$, $t = 3T$, etc. At time $t = 0$, the displacement x in the figure begins at $x(t = 0) = x_0 = 0$ with negative initial velocity $v(t = 0) = v_0 < 0$ and zero initial acceleration $a(t = 0) = 0$. As the mass is accelerated to the left, at time $t = T/4$ it reaches

$$x\left(t = \frac{T}{4}\right) = -\frac{v_0}{\omega} \cos\left(\frac{\pi}{2} + \frac{\pi}{2}\right)$$

$$= -\frac{v_0}{\omega} \cos(\pi)$$

$$= \frac{v_0}{\omega} \tag{11}$$

where use is made of $\cos(\pi) = \cos(180°) = -1$. Similarly, because $\sin(180°) = 0$, it can be shown that

$$v\left(t = \frac{T}{4}\right) = 0 \tag{12}$$

Because $v_0 < 0$, note that $x(t = T/4) = v_0/\omega < 0$. That is, at the quarter-time

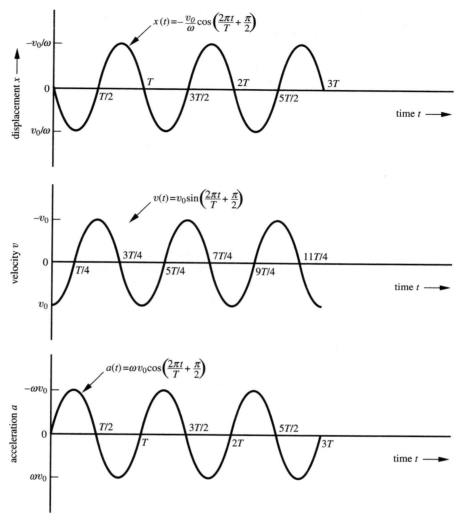

FIGURE 3.5.2 Graphical representation of the displacement x, velocity v, and acceleration a versus time t for the simple harmonic motion described by Equations (8), (9), and (10) in Example 3.5.2. Because $v_0 < 0$ is assumed, note that the maximum excursions $-v_0/\omega$, $-v_0$ and $-\omega v_0$ are all *positive* quantities in the figure.

period $t = T/4$, the mass is at the left-most extent of its oscillation. The mass then accelerates to the right, reaching the right-most extent of its oscillation at time $t = 3T/4$ as evident from the figure. The mass of course continues to oscillate back and forth with a total time period T required for each complete oscillation.

It is interesting to compare the motion in Figure 3.11, which corresponds to Equations (3.24), (3.25) and (3.26), with the motion considered in this exercise and illustrated in Figure 3.5.2. As expected, the motions are identical provided the time origin in Figure 3.11 is shifted by one-quarter time period *T*/4 *to the right.*

▶ **EXAMPLE 3.5.3**   A mass $m = 100$ g is attached to a spring and undergoes simple harmonic motion along a straight line according to

$$x(t) = 5 \cos\left(\frac{\pi t}{10} + \frac{\pi}{2}\right) \tag{1}$$

where the unit of displacement x is centimeters, and the unit of time t is seconds.

(a)   What is the oscillation period T?
(b)   What is the angular frequency of oscillation ω?
(c)   What is the oscillation amplitude A?
(d)   What is the phase angle ϕ?
(e)   What is the initial displacement $x(t = 0) = x_0$?
(f)   What is the expression for the velocity $v(t)$? What is the initial velocity $v(t = 0) = v_0$?
(g)   What is the expression for the acceleration $a(t)$?

In answer to parts (a)–(d), we compare the above expression for $x(t)$ with the general expression

$$x(t) = A \cos\left(\frac{2\pi t}{T} + \phi\right) \tag{2}$$

given in Equation (3.28). Equating the coefficients of time t in the two expressions, it is evident that the period T is determined from

$$\frac{\pi}{10} = \frac{2\pi}{T}$$

or

$$T = 2 \times 10$$
$$= 20 \text{ s} \tag{3}$$

Because $T = 2\pi/\omega$, we obtain for the angular frequency ω

$$\omega = \frac{2\pi}{T}$$

$$= \frac{2\pi}{20 \text{ s}}$$

$$= \frac{\pi}{10} \text{ rad/s} \tag{4}$$

Comparing the two expressions for $x(t)$, we also readily obtain

$$A = 5 \text{ cm} \tag{5}$$

for the oscillation amplitude A, and

$$\phi = \frac{\pi}{2} \tag{6}$$

for the phase angle ϕ.

In answer to part (e), we evaluate

$$x(t) = 5 \cos\left(\frac{\pi t}{10} + \frac{\pi}{2}\right) \tag{7}$$

at time $t = 0$. This gives

$$x(t = 0) = x_0$$

$$= 5 \cos\left(0 + \frac{\pi}{2}\right)$$

$$= 5 \cos\left(\frac{\pi}{2}\right)$$

$$= 0 \tag{8}$$

where use is made of $\cos(\pi/2) = \cos(90°) = 0$ (see Chapter 2).

In answer to part (f), we make use of Equation (3.29) with $T = 20$ s, $A = 5$ cm and $\phi = \pi/2$. This gives

$$v(t) = -\frac{\pi}{2} \sin\left(\frac{\pi t}{10} + \frac{\pi}{2}\right)$$

where the unit of velocity v is cm/s. To determine the initial velocity $v(t = 0) = v_0$, we set $t = 0$. This gives

$$v(t = 0) = v_0$$

$$= -\frac{\pi}{2} \sin\left(0 + \frac{\pi}{2}\right)$$

$$= -\frac{\pi}{2} \sin\left(\frac{\pi}{2}\right)$$

$$= -\frac{\pi}{2}$$

$$\simeq -1.57 \text{ cm/s} \tag{9}$$

where use is made of $\sin(\pi/2) = \sin(90°) = 1$.

Because $x_0 = 0$ and $v_0 = -1.57$ cm/s, it is evident that at time $t = 0$ the mass is passing through the origin $x_0 = 0$ and moving to the *left* with initial velocity $v_0 = -1.57$ cm/s.

In answer to part (g), we make use of Equation (3.30) with $T = 20$ s, $A = 5$ cm and $\phi = \pi/2$. This gives

$$a(t) = -\frac{\pi^2}{20} \cos\left(\frac{\pi t}{10} + \frac{\pi}{2}\right) \tag{10}$$

where the unit of acceleration a is cm/s^2. Note from the expression for $x(t)$ and $a(t)$ and $\omega = \pi/10$ rad/s that $a(t) = -\omega^2 x(t)$, as expected.

EXERCISES

1. For Example 3.5.1, draw graphs of the displacement, the velocity, and the acceleration as functions of time t from $t = 0$ to $t = 40$ s. (Compute the values at 5-s time intervals.)

2. Repeat for Example 3.5.2, with $T = 20$ s and $v_0 = -\pi/2$ cm/s $\simeq -1.57$ cm/s.

3. Construct a graph of

$$x(t) = 5 \sin(2\pi t)$$

as a function of time t from $t = 0$ to $t = 3$ s. Here, the displacement x is in centimeters. (Compute the values at 0.25-s time intervals.)

A mass $m = 10$ g is attached to a spring and undergoes simple harmonic motion along a straight line according to

$$x(t) = 10 \cos\left(2\pi t + \frac{\pi}{3}\right)$$

where the unit of displacement x is centimeters and the unit of time t is seconds. The notation is the same as in Equations (3.28), (3.29) and (3.30).

4. What is the oscillation period T?   (Ans. 353)

5. What is the angular frequency of oscillation ω?
(Ans. 38)

6. What is the spring constant k?   (Ans. 68)

7. What is the oscillation amplitude A? (Ans. 163)
8. What is the phase angle ϕ? (Ans. 273)
9. What is the initial displacement $x(t = 0) = x_0$?
   (Ans. 196)
10. What is the velocity $v(t)$? (Ans. 279)
11. What is the initial velocity $v(t = 0) = v_0$?
   (Ans. 261)
12. What is the acceleration $a(t)$? (Ans. 224)
13. Draw a graph of the displacement $x(t)$ as a function of time t from $t = 0$ to $t = 2$ s. (Compute the values in 0.25-s time intervals.)

A mass $m = 0.1$ g is attached to a spring with spring constant $k = 98.69$ g/s^2 $\simeq 10\pi^2$ g/s^2. The initial displacement of the mass is $x(t = 0) = x_0 = 7$ cm, and the initial velocity is $v(t = 0) = v_0 = 0$.

14. What is the angular frequency of oscillation ω?
   (Ans. 255)
15. What is the oscillation period T? (Ans. 206)
16. What is the displacement $x(t)$ as a function of time t?
   (Ans. 239)
17. What is the velocity $v(t)$? (Ans. 79)
18. What is the acceleration $a(t)$? (Ans. 300)
19. What is the maximum speed of the mass?
   (Ans. 337)
20. Draw a graph of the displacement $x(t)$ as a function of time t from $t = 0$ to $t = 0.4$ s. (Compute the values in 0.05-s time intervals.)

4

Derivatives

4.1 INSTANTANEOUS VELOCITY

In Section 3.4 we discussed the fact that the *slope* of the *x versus t* graph, which describes the motion of a particle in one dimension, is equal to the velocity of the particle:

$$v = \frac{\Delta x}{\Delta t} \tag{4.1}$$

The interpretation of the slope $\Delta x/\Delta t$ as the velocity is particularly simple when the *x versus t* graph is a straight line, as, for example, in Figure 3.2. But how do we calculate the velocity in the event that the *x versus t* graph is a *curve* instead of a straight line? A simple example will illustrate the difference in the two cases.

▶ **EXAMPLE 4.1.1**   A particle moves according to the relation

$$x(t) = \frac{1}{2} at^2 \tag{1}$$

where a is the constant acceleration (compare Equation 3.17). If the particle is falling freely near the surface of the earth, then $a = g = 32$ ft/s^2. Therefore, we have

$$x(t) = 16t^2 \tag{2}$$

where t is measured in seconds. What is the velocity of the particle at time $t = 5$ s?

Because the x *versus* t graph in this case is not a straight line, the calculation of the velocity according to Equation (4.1) actually yields the *average* velocity during the time interval Δt. During the time interval from $t = t_1$ to $t = t_2$ the average velocity is

$$\bar{v} = \frac{\Delta x}{\Delta t} = \frac{x_2 - x_1}{t_2 - t_1} \tag{3}$$

In (3), $\Delta t = t_2 - t_1$, and $\Delta x = x_2 - x_1$, where $x_2 = 16t_2^2$ is the displacement at time t_2 and $x_1 = 16t_1^2$ is the displacement at time t_1 (see the figure).

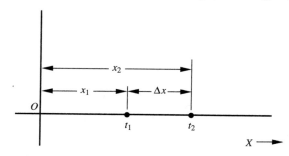

Let us compute $\bar{v}$ for several different intervals Δt, starting always at $t_1 = 5$ s, and $x_1 = 16(5)^2 = 400$ ft.

(a) $\Delta t = 1$ s: $t_1 = 5$ s, $t_2 = 6$ s

$$\bar{v} = \frac{16t_2^2 - 16t_1^2}{t_2 - t_1} = \frac{16(6)^2 - 16(5)^2}{6 - 5} = \frac{576 - 400}{1}$$

$$= 176 \text{ ft/s}$$

(b) $\Delta t = 0.01$ s: $t_1 = 5$ s, $t_2 = 5.01$ s

$$\bar{v} = \frac{16(5.01)^2 - 16(5)^2}{5.01 - 5} = \frac{401.6016 - 400}{0.01}$$

$$= 160.16 \text{ ft/s}$$

Clearly, we can calculate $\bar{v}$ in this way for any value of Δt. Some of the results of such calculations are given in the table below for $t_1 = 5$ s, and $x_1 = 16(5)^2 = 400$ ft.

$\Delta t = t_2 - t_1$ (s)	$\Delta x = x_2 - x_1$ (ft)	$\bar{v} = \dfrac{\Delta x}{\Delta t}$ (ft/s)
1	176	176
0.5	84.0	168
0.1	16.16	161.6
0.01	1.601 6	160.16
0.001	0.160 016	160.016
0.000 001	0.000 160 000 016	160.000 016

It is apparent from these results that $\bar{v}$ depends on the size of the time interval Δt. As Δt becomes smaller and smaller, Δx also becomes smaller and smaller. However, the ratio $\bar{v} = \Delta x / \Delta t$ appears to approach a constant value of 160 ft/s as Δt approaches closer and closer to zero.

The quantities we have calculated are the *average* velocities. But how do we find the velocity *exactly* at $t = 5$ s? What is the *instantaneous* velocity at this time? In order to answer these questions, we must use the methods of differential calculus.

4.2   LIMITS AND DERIVATIVES

Evaluating the *derivative* (or the *differentiation*) of a known function $f(t)$ is a very important operation in many areas of physics. In this section we first define the derivative of a function and show that it has a simple graphical interpretation. We then consider some examples and discuss some general properties of derivatives.

Consider the function $f(t)$ illustrated graphically by the solid curve in Figure 4.1. Note that t is the *independent* variable, and $f(t)$ is the *dependent* variable. The function $f(t)$ could represent (for example) the instantaneous velocity $v(t)$ or displacement $x(t)$ of a particle at time t. For present purposes, however, it is not necessary to specify

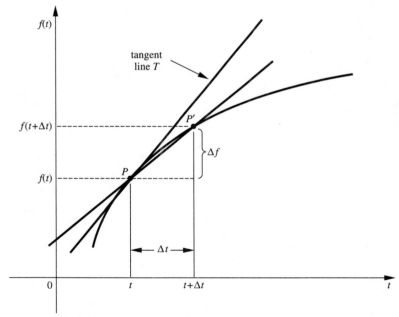

FIGURE 4.1   In the limit that Δt becomes very small, the slope of PP' approaches that of the tangent line T.

the physical significance *or* functional form of $f(t)$. Nor is it necessary to specify the units of t.

Let us compare the values of the function at two different times, $t + \Delta t$ and t. From Figure 4.1, the difference between $f(t + \Delta t)$ and $f(t)$ is

$$\Delta f = f(t + \Delta t) - f(t) \tag{4.2}$$

Moreover, the *slope* of the straight line passing through the two points P and P' is (see Section 3.3)

$$\text{Slope} = \frac{\Delta f}{\Delta t} = \frac{f(t + \Delta t) - f(t)}{\Delta t} \tag{4.3}$$

The derivative of $f(t)$ is closely related to the slope defined in Equation (4.3). By convention, the *derivative of* $f(t)$ *with respect to* t is denoted by

$$\frac{df}{dt}$$

This symbol does *not* mean "*df* divided by *dt*."

By definition, the derivative of f(t) with respect to t is equal to the value of the slope of the line passing through P and P' in the limit that Δt *approaches zero* (see Figure 4.1). That is, df/dt is defined to be

$$\frac{df}{dt} = \lim_{\Delta t \to 0} \frac{\Delta f}{\Delta t} = \lim_{\Delta t \to 0} \frac{f(t + \Delta t) - f(t)}{\Delta t} \tag{4.4}$$

The notation

$$\lim_{\Delta t \to 0}$$

used in Equation (4.4) signifies *in the limit that* Δt *approaches zero*. Operationally it means that Δt becomes very, very small, that is, *arbitrarily close to zero*.

As Δt becomes smaller and smaller, note from Figure 4.1 that P' approaches P, Δf approaches zero, and the line through P and P' approaches the line T, which is tangent to the curve at the point P.

In the limit that Δt becomes arbitrarily small, the line through P and P' *coincides* with the tangent line T. Therefore, comparing Equations (4.4) and Figure 4.1, the derivative of $f(t)$ with respect to t can be interpreted as the *slope of the tangent line* to the $f(t)$ *versus* t curve:

$$\frac{df}{dt} = \text{slope of the } f(t) \text{ } versus \text{ } t \text{ curve} \tag{4.5}$$

Note that df/dt is generally a *variable function of* t because the slope of the $f(t)$ *versus* t curve is different for different values of t (see Figure 4.1).

Although Δf approaches zero as Δt approaches zero, the ratio $\Delta f/\Delta t$ tends to a *finite* value, at least for sufficiently ''smooth'' functions $f(t)$. We can illustrate this point by returning to the example of the falling particle (Example 4.1.1).

▶ **EXAMPLE 4.2.1**   A freely falling object obeys the equation

$$x(t) = 16t^2 \tag{1}$$

In order to calculate the *instantaneous velocity* at time $t = t_1$, we set $t_2 = t_1 + \Delta t$. Then,

$$\overline{v} = \frac{\Delta x}{\Delta t} = \frac{x(t_2) - x(t_1)}{\Delta t} = \frac{x(t_1 + \Delta t) - x(t_1)}{\Delta t} \tag{2}$$

Using (1),

$$\overline{v} = \frac{16(t_1 + \Delta t)^2 - 16(t_1)^2}{\Delta t}$$

$$= \frac{16t_1^2 + 32t_1\Delta t + 16(\Delta t)^2 - 16t_1^2}{\Delta t}$$

$$= \frac{32t_1\Delta t + 16(\Delta t)^2}{\Delta t}$$

$$= 32t_1 + 16\Delta t \tag{3}$$

Now, we let $\Delta t \to 0$. The second term in (3) then vanishes and we have

$$v = 32t_1 \tag{4}$$

We have removed the bar denoting the average value because in the limit $\Delta t \to 0$, $\overline{v}, \to v$, the *instantaneous* velocity. For $t_1 = 5$ s, we find

$$v = 160 \text{ ft/s} \tag{5}$$

and we see that $v(t = 5$ s$)$ does, in fact, equal the value that $\overline{v}$ seemed to be approaching in Example 4.1.1.

Notice that in the example above we have calculated the derivative of x with respect to t. [Compare (2) with Equation 4.4.] That is, we have shown that the *instantaneous velocity* is given by

$$v(t) = \lim_{\Delta t \to 0} \frac{x(t + \Delta t) - x(t)}{\Delta t}$$

or

$$v(t) = \frac{dx}{dt} \tag{4.6}$$

In Equation (4.6) we write $v(t)$ to indicate explicitly that the instantaneous velocity depends, in the general case, upon the time t.

EXAMPLE 4.2.2   Calculate df/dt for $f(t) = t^3$.

For this example,

$$f(t + \Delta t) = (t + \Delta t)^3 = t^3 + 3t^2\Delta t + 3t(\Delta t)^2 + (\Delta t)^3 \tag{1}$$

where use has been made of the *binomial expansion* of $(t + \Delta t)^3$, given in Equation (1.32). Therefore,

$$\frac{df}{dt} = \lim_{\Delta t \to 0} \frac{f(t + \Delta t) - f(t)}{\Delta t}$$

$$= \lim_{\Delta t \to 0} \frac{t^3 + 3t^2\Delta t + 3t(\Delta t)^2 + (\Delta t)^3 - t^3}{\Delta t} \tag{2}$$

In the limit that Δt approaches zero, the terms in (2) proportional to $(\Delta t)^2/\Delta t$ and $(\Delta t)^3/\Delta t$ vanish. This leaves only the term $3t^2\Delta t/\Delta t$, whose limit is equal to $3t^2$.

Therefore, for $f(t) = t^3$,

$$\frac{df}{dt} = 3t^2 \tag{3}$$

The result in (3) can also be expressed as

$$\frac{d(t^3)}{dt} = \frac{d}{dt}(t^3) = 3t^2 \tag{4}$$

Note for this example that df/dt is a *variable function* of t.

The analysis in Example 4.2.2 can be extended to the case that $f(t) = t^n$, where n is any constant. Making use of the definition of the derivative, and the binomial expansion for $(t + \Delta t)^n$, it is easy to show

$$\frac{d}{dt}(t^n) = nt^{n-1} \tag{4.7}$$

Equation (4.7) and examples of other derivatives often encountered in physics are

TABLE 4.1   Examples of Derivatives

$f(t)$	$\dfrac{df}{dt}$
$\alpha = \text{constant}$	0
t	1
t^2	$2t$
t^3	$3t^2$
$\sqrt{t}$	$1/(2\sqrt{t})$
$1/t$	$-1/t^2$
$1/t^2$	$-2/t^3$
$1/t^3$	$-3/t^4$
$1/\sqrt{t}$	$-1/(2t^{3/2})$
t^n	nt^{n-1}
$\sin t$	$\cos t$
$\sin \alpha t$	$\alpha \cos \alpha t$
$\cos t$	$-\sin t$
$\cos \alpha t$	$-\alpha \sin \alpha t$
$\tan t$	$1/\cos^2 t = \sec^2 t$
$\tan \alpha t$	$\alpha/\cos^2 \alpha t = \alpha \sec^2 \alpha t$
e^t	e^t
$e^{\alpha t}$	$\alpha e^{\alpha t}$
e^{-t}	$-e^{-t}$
$e^{-\alpha t}$	$-\alpha e^{-\alpha t}$
$\ln t$	$1/t$
$\ln \alpha t$	$1/t$

summarized in Table 4.1. Note that this table includes derivatives of trigonometric, exponential, and logarithmic functions. These will be discussed further in Sections 4.4–4.6

A seemingly trivial (but nonetheless important) point should be noted. Suppose the independent variable is x (rather than t) and the dependent variable is $g(x)$. Then the derivative of $g(x)$ with respect to x is defined in a manner completely analogous to Equation (4.4):

$$\frac{dg}{dx} = \lim_{\Delta x \to 0} \frac{g(x + \Delta x) - g(x)}{\Delta x}$$

Comparing with Table 4.1, we find $d(x^3)/dx = 3x^2$, $d(x^n)/dx = nx^{n-1}$, and so on.

In concluding this section we summarize several important rules that characterize the operation of differentiation.

The derivative of a constant times a function is equal to the constant times the derivative of the function. That is,

$$\frac{d}{dt}[\alpha f(t)] = \alpha \frac{df}{dt} \tag{4.8}$$

where α is a constant and $f(t)$ is an arbitrary function.

The derivative of the sum of any two functions is equal to the sum of their derivatives. That is,

$$\frac{d}{dt}[\alpha f(t) + \beta g(t)] = \alpha \frac{df}{dt} + \beta \frac{dg}{dt} \qquad \text{(4.9)}$$

where α and β are constants, and $f(t)$ and $g(t)$ are arbitrary functions. Equation (4.9) is called the *linearity* property of differentiation.

The derivative of the *product* of any two functions, $f(t)$ and $g(t)$, can be expressed as

$$\frac{d}{dt}(fg) = f\frac{dg}{dt} + g\frac{df}{dt} \qquad \text{(4.10)}$$

To derive the result in Equation (4.10) we make use of the definition of the derivative:

$$\frac{d}{dt}(fg) = \lim_{\Delta t \to 0} \frac{f(t + \Delta t)\, g(t + \Delta t) - f(t)\, g(t)}{\Delta t} \qquad \text{(4.11)}$$

Adding and subtracting $f(t + \Delta t)\, g(t)/\Delta t$ on the right-hand side of Equation (4.11), and rearranging terms, gives

$$\frac{d}{dt}(fg) = \lim_{\Delta t \to 0} \left[f(t + \Delta t)\left\{ \frac{g(t + \Delta t) - g(t)}{\Delta t} \right\} + g(t)\left\{ \frac{f(t + \Delta t) - f(t)}{\Delta t} \right\} \right] \qquad \text{(4.12)}$$

In the limit that Δt approaches zero, the terms in curly brackets approach dg/dt and df/dt, respectively, whereas the coefficient $f(t + \Delta t)$ approaches $f(t)$. Therefore, Equation (4.10) follows directly from Equation (4.12).

The derivative of the *quotient* of any two functions, $f(t)$ and $g(t)$, can be expressed as

$$\frac{d}{dt}\left(\frac{f}{g}\right) = \frac{1}{g^2}\left(g\frac{df}{dt} - f\frac{dg}{dt} \right) \qquad \text{(4.13)}$$

The proof of this result is left as an exercise.

▶ **EXAMPLE 4.2.3**   Verify that Equation (4.13) is satisfied for $f(t) = t^3$ and $g(t) = t$.

Using Equation (4.13), we have

$$\frac{d}{dt}\left(\frac{f}{g}\right) = \frac{1}{t^2}\left(t\frac{dt^3}{dt} - t^3\frac{dt}{dt} \right) \qquad \text{(1)}$$

From Table 4.1,

$$\frac{dt^3}{dt} = 3t^2 \, ; \quad \frac{dt}{dt} = 1 \tag{2}$$

Therefore,

$$\frac{d}{dt}\left(\frac{f}{g}\right) = \frac{1}{t^2}(t \times 3t^2 - t^3 \times 1)$$

$$= \frac{1}{t^2}(3t^3 - t^3)$$

$$= 2t \tag{3}$$

Notice in this case, we can divide $f(t)$ by $g(t)$ *before* differentiating, obtaining

$$\frac{d}{dt}\left(\frac{f}{g}\right) = \frac{d}{dt}\left(\frac{t^3}{t}\right) = \frac{d}{dt}(t^2) = 2t \tag{4}$$

and (4) agrees with the result found by using the rule for differentiating a quotient.

One of the important rules of differentiation is known as the *chain rule*. Assume that g is a function of f [$g = g(f)$], and that f is a function of t [$f = f(t)$]. Since g varies with f, and f varies with t, it follows that g *also* varies with t. Without presenting the proof, it can be shown that

$$\frac{dg}{dt} = \frac{dg}{df} \cdot \frac{df}{dt} \tag{4.14}$$

This equation is known as the *chain rule of differentiation*. It states that dg/dt is the product of the two derivatives, dg/df and df/dt.

▶ **EXAMPLE 4.2.4**   The kinetic energy of a particle as a function of velocity is

$$E_K(v) = \frac{1}{2}mv^2 \tag{1}$$

Evaluate dE_K/dt if the particle's velocity varies with time according to

$$v(t) = bt^4 \tag{2}$$

where b is a constant.
    From the chain rule,

$$\frac{dE_K}{dt} = \frac{dE_K}{dv} \cdot \frac{dv}{dt} \tag{3}$$

From (1) and (2), and Table 4.1, it follows that

$$\frac{dE_K}{dv} = \frac{d}{dv}\left(\frac{1}{2}mv^2\right) = \frac{1}{2}m(2v) = mv \tag{4}$$

and

$$\frac{dv}{dt} = \frac{d}{dt}(bt^4) = 4bt^3 \tag{5}$$

Combining (3), (4), and (5) gives

$$\frac{dE_K}{dt} = (mv) \times (4bt^3) \tag{6}$$

Since $v(t) = bt^4$, (6) can be expressed as

$$\frac{dE_K}{dt} = (mbt^4) \times (4bt^3)$$

$$= 4mb^2t^7 \tag{7}$$

Equation (7) can also be obtained *without* the chain rule. Substituting (2) into (1) gives E_K directly as a function of t:

$$E_K(t) = \frac{1}{2}m(bt^4)^2 = \frac{1}{2}mb^2t^8 \tag{8}$$

Making use of Table 4.1, we find

$$\frac{dE_K}{dt} = \frac{d}{dt}\left(\frac{1}{2}mb^2t^8\right)$$

$$= 4mb^2t^7 \tag{9}$$

which is identical to the result obtained in (7) by means of the chain rule.

▶ **EXAMPLE 4.2.5** Calculate the derivative df/dt from first principles for the case where

$$f(t) = t^n \tag{1}$$

From Equation (4.4), the derivative df/dt is defined by

$$\frac{df}{dt} = \lim_{\Delta t \to 0} \frac{f(t + \Delta t) - f(t)}{\Delta t} \tag{2}$$

Making use of $f = t^n$ gives

$$\frac{d}{dt}(t^n) = \lim_{\Delta t \to 0} \frac{(t + \Delta t)^n - t^n}{\Delta t} \tag{3}$$

To simplify (3), we make use of the binomial expansion

$$(1 + x)^n = 1 + nx + \frac{n(n - 1)}{2}x^2 + \cdots \tag{4}$$

in Equation (1.31). We also rewrite the expression for $(t + \Delta t)^n$ in the form

$$(t + \Delta t)^n = t^n\left(1 + \frac{\Delta t}{t}\right)^n \tag{5}$$

Comparing Equations (4) and (5), we obtain

$$\left(1 + \frac{\Delta t}{t}\right)^n = 1 + n\frac{\Delta t}{t} + \frac{n(n - 1)}{2}\left(\frac{\Delta t}{t}\right)^2 + \cdots \tag{6}$$

for small values of $\Delta t/t$, where the identification $x = \Delta t/t$ has been made. Substituting Equations (5) and (6) into Equation (3) gives

$$\frac{d}{dt}(t^n) = \lim_{\Delta t \to 0} \left\{\frac{t^n\left[1 + n\frac{\Delta t}{t} + \frac{n(n - 1)}{2}\left(\frac{\Delta t}{t}\right)^2 + \cdots\right] - t^n}{\Delta t}\right\}$$

$$= \lim_{\Delta t \to 0} \left\{\frac{nt^n\Delta t}{\Delta t\, t} + \frac{n(n - 1)t^n}{2\Delta t}\left(\frac{\Delta t}{t}\right)^2 + \cdots\right\} \tag{7}$$

The second term on the right-hand side of Equation (7) vanishes in the limit $\Delta t \to 0$, which gives

$$\frac{d}{dt}(t^n) = \frac{nt^n}{t} \tag{8}$$

or

$$\frac{d}{dt}(t^n) = nt^{n-1}$$

which is the desired result (see Table 4.1).

EXERCISES   *Using the basic definition of the derivative in Equation (4.4), verify directly that:*

1. $\dfrac{d}{dt}(b) = 0$, for b = constant.

2. $\dfrac{d}{dx}(x^2) = 2x$.

3. $\dfrac{d}{dt}(t^4) = 4t^3$.

4. $\dfrac{d}{dx}\left(\dfrac{1}{x}\right) = -\dfrac{1}{x^2}$.

5. $\dfrac{d}{dt}\left(\dfrac{1}{t^2}\right) = -\dfrac{2}{t^3}$.

*Using Table **4.1**, verify that the rule for differentiating the quotient of two functions (Equation **4.13**) is satisfied for:*

6. $f(t) = t^6$, and $g(t) = t^n$.

7. $f(t) = 1/t^2$, and $g(t) = t^2$.

8. $f(t) = t^4$, and $g(t) = t$.

9. $f(t) = 0$, and $g(t) = t^{963}$.

*Use Equation **(4.13)** to verify the following:*

10. $\dfrac{d}{dt}\left(\dfrac{1}{g}\right) = -\dfrac{1}{g^2}\dfrac{dg}{dt}$.

11. $\dfrac{d}{dt}\left(\dfrac{1}{t}\right) = -\dfrac{1}{t^2}$.

12. $\dfrac{d}{dt}\left(\dfrac{1}{t^2}\right) = -\dfrac{2}{t^3}$.

13. Consider the function $f(t) = t^n$ for $t \geqslant 0$, and arbitrary n. Verify that $df/dt \geqslant f/t$ for $n \geqslant 1$.

Use the chain rule to evaluate dg/dt for the following choices of g(f) and f(t):

14. $g(f) = f^4$, and $f(t) = t^3$ (Ans. 231)

15. $g(f) = 1/f^3$, and $f(t) = 3t^{10}$ (Ans. 377)

16. $g(f) = f^{3/2}$, and $f(t) = t^2$ (Ans. 367)

17. $g(f) = f^3$, and $f(t) = 1/t^2$ (Ans. 180)

4.3   DERIVATIVE RELATIONS FOR ONE-DIMENSIONAL MOTION

We have already seen that the instantaneous velocity is equal to the derivative of $x(t)$ with respect to t.

$$v(t) = \frac{dx}{dt} \tag{4.15}$$

where we write $v(t)$ to indicate explicitly that the velocity depends, in the general case, upon the time t.

If the velocity of an object is changing (either increasing or decreasing), we say that the object is undergoing acceleration. The *average acceleration* $\bar{a}$ during a time interval Δt is defined in a way analogous to that for average velocity. We write

$$\bar{a} = \frac{\Delta v}{\Delta t} \tag{4.16}$$

where Δv is the change in velocity during the time interval Δt. If the acceleration is constant, the velocity-time graph is a straight line and the slope is equal to $\bar{a}$.

Generally, however, the acceleration can depend on the time, so we express the *instantaneous acceleration a(t)* in the form

$$a(t) = \lim_{\Delta t \to 0} \frac{v(t + \Delta t) - v(t)}{\Delta t}$$

or

$$a(t) = \frac{dv}{dt} \qquad \text{(4.17)}$$

Because $v(t) = dx/dt$, the acceleration can also be expressed as

$$a(t) = \frac{d}{dt}\left(\frac{dx}{dt}\right) \qquad \text{(4.18)}$$

The right-hand side of this equation signifies ''the derivative of dx/dt with respect to t.'' The notation in Equation (4.18) is often contracted to

$$a(t) = \frac{d^2x}{dt^2} \qquad \text{(4.19)}$$

Whereas dx/dt is called the *first* derivative of $x(t)$ with respect to t, d^2x/dt^2 is called the *second* derivative of $x(t)$ with respect to t. Operationally, d^2x/dt^2 signifies: *first* evaluate dx/dt, and *then* evaluate the derivative of dx/dt with respect to t.

▶ **EXAMPLE 4.3.1**   If $x(t) = \alpha t^m$, where α and m are constants, calculate $a(t)$ using Equation (4.19). We have

$$a(t) = \frac{d^2x}{dt^2} = \frac{d}{dt}\left(\frac{d}{dt}\,\alpha t^m\right)$$

$$= \frac{d}{dt}\,(m\alpha t^{m-1})$$

$$= m(m-1)\alpha t^{m-2}$$

where we have used

$$\frac{d}{dt}\,(t^n) = nt^{n-1}$$

first for $n = m$ and then for $n = m - 1$.

The general equation for the position of a particle undergoing one-dimensional motion subject to a constant force F_0 is

$$x(t) = x_0 + v_0 t + \frac{1}{2}\frac{F_0}{m} t^2 \tag{4.20}$$

By differentiating this expression for $x(t)$ we can obtain the general result for $v(t)$:

$$v(t) = \frac{dx}{dt} = \frac{d}{dt}\left(x_0 + v_0 t + \frac{1}{2}\frac{F_0}{m} t^2\right)$$

$$= v_0 + \frac{F_0}{m} t \tag{4.21}$$

A further differentiation yields the acceleration:

$$a(t) = \frac{d^2 x}{dt^2} = \frac{dv}{dt} = \frac{d}{dt}\left(v_0 + \frac{F_0}{m} t\right)$$

$$= \frac{F_0}{m} = a_0 \tag{4.22}$$

which shows that acceleration is constant (denoted by a_0). Therefore, the results can be summarized as

$$x(t) = x_0 + v_0 t + \frac{1}{2} a_0 t^2$$

$$v(t) = v_0 + a_0 t \tag{4.23}$$

$$a(t) = a_0 = \text{constant}$$

It is important to keep in mind the geometrical significance of these relations. If we simplify the situation by setting $x_0 = 0$ and $v_0 = 0$ (that is, the particle starts from rest at $x = 0$ when $t = 0$), then

$$\text{displacement} = x(t) = \frac{1}{2} a_0 t^2 \tag{4.24a}$$

$$\text{velocity} = v(t) = a_0 t \tag{4.24b}$$

$$\text{acceleration} = a_0 \tag{4.24c}$$

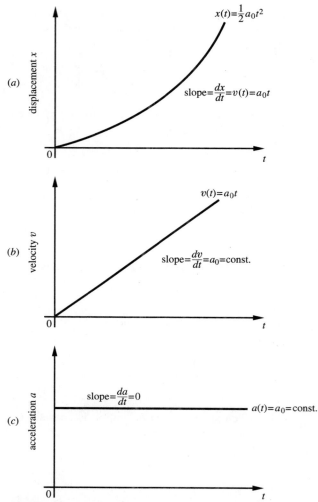

FIGURE 4.2 Graphical representations of (a) $x(t)$, (b) $v(t)$, and
(c) $a(t)$ for uniformly accelerated motion. Note the
significance of the slope of the graph in each case.

These functions are illustrated graphically in Figure 4.2 where it is emphasized that
the slope of the x-t curve gives the velocity, and the slope of the v-t curve gives
the acceleration.

EXERCISES

*The displacement of an object of mass $m = 5$ kg is observed to increase with time according
to $x(t) = 32t^5$, where t is in seconds and $x(t)$ is in meters. Calculate the following properties
of the object's motion.*

1. The average velocity $\bar{v}$ for the time interval from $t = 2$ s to $t = 3$ s.   (Ans. 366)
2. The instantaneous velocity at $t = 2$ s.   (Ans. 161)
3. The instantaneous velocity at $t = 2t_0$.   (Ans. 238)
4. The average acceleration $\bar{a}$ for the time interval from $t = 2$ s to $t = 3$ s.   (Ans. 27)
5. The instantaneous acceleration at $t = 2$ s.   (Ans. 140)

6. The instantaneous acceleration at $t = 2t_0$.   (Ans. 188)
7. The force on the object at time t.   (Ans. 383)
8. Is the motion uniformly accelerated motion?   (Ans. 4)

Evaluate d^2f/dt^2 for the following functions $f(t)$:

9. $f(t) = t^{11/3}$ (Ans. 307)
10. $f(t) = 1/t^3$ (Ans. 259)
11. $f(t) = t$ (Ans. 210)

4.4 DERIVATIVES OF TRIGONOMETRIC FUNCTIONS

In Sections 2.7 to 2.12, we examined several general properties of the trigonometric functions sin θ, cos θ, and tan θ. These functions are illustrated graphically in Figures 2.24 to 2.26. In this section, we discuss the *derivatives* of trigonometric functions.

Consider a general function $f(\theta)$, in which θ (the *independent* variable) is in *radians*. As in Equation (4.4), in which t is the independent variable, the derivative of $f(\theta)$ with respect to θ is defined by

$$\frac{df}{d\theta} = \lim_{\Delta\theta \to 0} \frac{f(\theta + \Delta\theta) - f(\theta)}{\Delta\theta} \tag{4.25}$$

Moreover, $df/d\theta$ is equal to the *slope* of the $f(\theta)$ *versus* θ curve.

Shown in Figure 4.3 are graphs of sin θ, cos θ, and $-$ sin θ, for θ in the range from 0 to 3π radians. A careful examination of Figure 4.3 shows that for each value of θ the slope of the sin θ curve is equal to cos θ, and the slope of the cos θ curve is equal to $-$ sin θ. For example, if $\theta = \pi/2$ radians, the slope of sin $\theta = 0 =$ cos $\pi/2$, and the slope of cos $\theta = -1 = -$ sin $\pi/2$.

For any value of θ it can be shown that

$$\frac{d}{d\theta}(\sin \theta) = \cos \theta \tag{4.26}$$

$$\frac{d}{d\theta}(\cos \theta) = -\sin \theta \tag{4.27}$$

It is indeed remarkable that the derivative of a trigonometric function is itself equal to a trigonometric function.

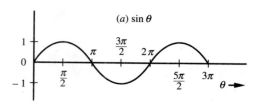

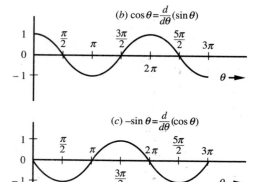

FIGURE 4.3   Graphs of (a) sin θ, (b) cos θ, and (c) $-\sin \theta$ *versus* the angle θ. The functions sin θ and cos θ are related by cos $\theta = d \sin \theta/d\theta$ and $-\sin \theta = d \cos \theta/d\theta$.

Equations (4.26) and (4.27) also give information concerning the *second* derivative of sin θ and cos θ (see Equation 4.19). Substituting cos $\theta = d \sin \theta/d\theta$ in Equation (4.27) gives

$$\frac{d^2}{d\theta^2} (\sin \theta) = -\sin \theta \tag{4.28}$$

Similarly, substituting sin $\theta = -d \cos \theta/d\theta$ in Equation (4.26) gives

$$\frac{d^2}{d\theta^2} (\cos \theta) = -\cos \theta \tag{4.29}$$

In each case the second derivative of the function is equal to the *negative* of the same function.

The derivative of tan θ with respect to θ can be evaluated as follows. First note that from the definition of tan θ (see Equation 2.18), tan $\theta = \sin \theta/\cos \theta$. Therefore,

$$\frac{d}{d\theta} (\tan \theta) = \frac{d}{d\theta} \left(\frac{\sin \theta}{\cos \theta} \right)$$

TABLE 4.2   Derivatives of Trigonometric Functions

$f(t)$	$\dfrac{df}{dt}$
$\sin t$	$\cos t$
$\sin \alpha t$	$\alpha \sin \alpha t$
$\cos t$	$-\sin t$
$\cos \alpha t$	$-\alpha \sin \alpha t$
$\tan t$	$1/\cos^2 t = \sec^2 t$
$\tan \alpha t$	$\alpha/\cos^2 \alpha t = \alpha\sec^2 \alpha t$

Making use of the rule for differentiating a quotient (Equation 4.13), we find

$$\frac{d}{d\theta}(\tan \theta) = \frac{1}{\cos^2 \theta}\left[\cos \theta \frac{d}{d\theta}(\sin \theta) - \sin \theta \frac{d}{d\theta}(\cos \theta)\right]$$

Combining this result with Equations (4.26) and (4.27) gives

$$\frac{d}{d\theta}(\tan \theta) = \frac{1}{\cos^2 \theta}[\cos \theta \times \cos \theta - \sin \theta \times (-\sin \theta)]$$

$$= \frac{1}{\cos^2 \theta}[\cos^2 \theta + \sin^2 \theta]$$

Since $\cos^2 \theta + \sin^2 \theta = 1$, and $\sec^2 \theta = 1/\cos^2 \theta$, we obtain, finally,

$$\frac{d}{d\theta}(\tan \theta) = \sec^2 \theta \qquad \text{(4.30)}$$

Equations (4.26) to (4.30) constitute very important derivative properties of trigonometric functions.

In many applications in physics, we encounter trigonometric functions of the form (for example)

$$\sin \alpha t$$

where t is the *independent variable* (e.g., *time*), and α is a constant. What is the derivative of $\sin \alpha t$ with respect to t? The answer to this question is readily obtained from the *chain rule* for differentiation (Equation 4.14). The reader should carry out these operations and verify the derivatives for $\sin \alpha t$, $\cos \alpha t$, and $\tan \alpha t$ shown in Table 4.2 (see also Table 4.1).

► **EXAMPLE 4.4.1**   In Section 3.5, we discussed the equations that describe the simple harmonic motion of a mass m along a straight line. For a mass attached to a spring with spring constant k, the displacement $x(t)$ as a function of time t can be expressed from Equation (3.28) as

$$x(t) = A \cos\left(\frac{2\pi t}{T} + \phi\right) \tag{1}$$

Here, the constants A and ϕ are the oscillation amplitude and phase, respectively, $T = 2\pi/\omega$ is the period of oscillation, and $\omega = \sqrt{k/m}$ is the angular frequency of oscillation.

(a)   From the definition of instantaneous velocity $v(t)$ in Equation (4.6), make use of Equation (1) to determine $v(t)$ from first principles.

(b)   Make use of Table 4.1 and the answer in part (a) to calculate the instantaneous acceleration $a(t)$.

In answer to part (a), the instantaneous velocity is

$$v(t) = \frac{dx}{dt}$$

$$= \lim_{\Delta t \to 0} \frac{x(t + \Delta t) - x(t)}{\Delta t} \tag{2}$$

where use is made of Equation (4.6). Substituting Equation (1) into Equation (2) gives

$$v(t) = \lim_{\Delta t \to 0} \frac{A \cos\left[\dfrac{2\pi(t + \Delta t)}{T} + \phi\right] - A \cos\left(\dfrac{2\pi t}{T} + \phi\right)}{\Delta t} \tag{3}$$

We make use of the cosine summation rule in Equation (2.23) to express

$$\cos(\alpha + \beta) = \cos\alpha \cos\beta - \sin\alpha \sin\beta \tag{4}$$

Substituting

$$\alpha = \frac{2\pi t}{T} + \phi$$

$$\beta = \frac{2\pi\Delta t}{T} \tag{5}$$

in Equation (4), we obtain

$$\cos\left[\frac{2\pi(t + \Delta t)}{T} + \phi\right] = \cos\left(\frac{2\pi t}{T} + \phi\right)\cos\left(\frac{2\pi\Delta t}{T}\right)$$

$$- \sin\left(\frac{2\pi t}{T} + \phi\right)\sin\left(\frac{2\pi\Delta t}{T}\right) \tag{6}$$

Making use of (6) in (3) gives

$$v(t) = \lim_{\Delta t \to 0} A\left\{\frac{1}{\Delta t}\cos\left(\frac{2\pi t}{T} + \phi\right)\left[\cos\left(\frac{2\pi\Delta t}{T}\right) - 1\right]\right.$$

$$\left. - \frac{1}{\Delta t}\sin\left(\frac{2\pi t}{T} + \phi\right)\sin\left(\frac{2\pi\Delta t}{T}\right)\right\} \tag{7}$$

In Equation (7), for small Δt, the difference

$$\cos\left(\frac{2\pi\Delta t}{T}\right) - 1 = 1 - \frac{1}{2}\left(\frac{2\pi\Delta t}{T}\right)^2 + \cdots - 1$$

$$\simeq -\frac{1}{2}\left(\frac{2\pi\Delta t}{T}\right)^2 \tag{8}$$

rapidly approaches zero proportional to $(\Delta t)^2$ as $\Delta t \to 0$. On the other hand, making use of the small-angle approximation $\sin\theta \simeq \theta$ in Equation (2.30), it follows that

$$\sin\left(\frac{2\pi\Delta t}{T}\right) \simeq \frac{2\pi\Delta t}{T} \tag{9}$$

for $\Delta t \to 0$. Substituting (8) and (9) into (7) readily gives for $\Delta t \to 0$

$$v(t) = \lim_{\Delta t \to 0} \frac{-A\sin\left(\frac{2\pi t}{T} + \phi\right)\frac{2\pi\Delta t}{T}}{\Delta t}$$

or equivalently

$$v(t) = -\frac{2\pi A}{T}\sin\left(\frac{2\pi t}{T} + \phi\right) \tag{10}$$

which is identical to Equation (3.29).

As an alternate derivation of Equation (10), we make use of Table 4.1 to express

$$\frac{d}{dt} \cos(\alpha t + \phi) = -\alpha \sin(\alpha t + \phi) \tag{11}$$

Combining Equations (1) and (11) with $\alpha = 2\pi/T$ gives

$$v(t) = \frac{dx}{dt}$$

$$= A \frac{d}{dt} \cos\left(\frac{2\pi t}{T} + \phi\right)$$

$$= -\frac{2\pi A}{T} \sin\left(\frac{2\pi t}{T} + \phi\right) \tag{12}$$

which is the required result.

In answer to part (b), we make use of Table 4.1 to express

$$\frac{d}{dt} \sin(\alpha t + \phi) = \alpha \cos(\alpha t + \phi) \tag{13}$$

Combining Equations (10) and (13), we obtain for the instantaneous acceleration

$$a(t) = \frac{dv}{dt}$$

$$= -\frac{2\pi A}{T} \frac{d}{dt} \sin\left(\frac{2\pi t}{T} + \phi\right)$$

$$= -\left(\frac{2\pi}{T}\right)^2 A \cos\left(\frac{2\pi t}{T} + \phi\right) \tag{14}$$

which is identical to Equation (3.30).

EXERCISES

Use Table 4.2 and the rule for differentiating a product (Equation 4.10) to evaluate the following derivatives:

1. $\dfrac{d}{dt} (\cos^2 t)$ (Ans. 378)

2. $\dfrac{d}{dt} (\sin^2 6t)$ (Ans. 348)

3. $\dfrac{d}{dt} (\sin t \cos 2t)$ (Ans. 83)

4. $\dfrac{d}{dt} (\cos t \tan t)$ (Ans. 342)

5. $\dfrac{d}{dt} (\cos t \sin^2 t)$ (Ans. 52)

Use Table **4.2** *and the rule for differentiating a quotient (Equation* **4.13**) *to evaluate the following derivatives:*

6. $\dfrac{d}{dt}\left(\dfrac{\sin t}{\cos t}\right) = \dfrac{d}{dt}(\tan t)$　　　　(Ans. 73)　　**8.** $\dfrac{d}{dt}\left(\dfrac{1}{\cos 3t}\right)$　　　　(Ans. 319)

7. $\dfrac{d}{dt}\left(\dfrac{\sin t}{\tan t}\right)$　　　　(Ans. 264)

Use Table **4.1** *and the rule for differentiating a product (Equation* **4.10**) *to evaluate the following derivatives:*

9. $\dfrac{d}{dt}(t^4 \sin 6t)$　　　　(Ans. 96)　　**10.** $\dfrac{d}{dt}(t^n \tan \alpha t)$　　　　(Ans. 333)

4.5　EXPONENTIAL FUNCTIONS AND THEIR DERIVATIVES

The function

$$g(t) = t^a \qquad \textbf{(4.31)}$$

where a is a constant, is called an *algebraic function* of the variable t. The function

$$f(t) = a^t \qquad \textbf{(4.32)}$$

in which the variable appears as an exponent, is called an *exponential function* of t. The constant a in Equation 4.32 is called the *base number* of the function.

In all kinds of scientific work, there frequently occur exponential functions in which the base number, denoted by e, is a certain irrational number,

$$e \cong 2.7183 \qquad \textbf{(4.33)}$$

(More accurately, $e = 2.7182818284\dots$.) The exponential function with e as the base number is written as

$$f(t) = e^t \qquad \textbf{(4.34)}$$

The function e^t is called the *standard exponential function*, although the term "standard" is often omitted. Sometimes we see the exponential function written as $f(t) = \exp t$.

The function e^t is characterized by a very *rapid* increase as t increases. This is

TABLE 4.3	Exponential Function to the Base e
t	e^t
0	1.0000
2	7.3891
4	54.598
6	403.43
8	2981.0
10	22026.

illustrated in Table 4.3, where the values of e^t are tabulated for t ranging from 0 to 10 in steps of 2. For example, at $t = 0$, the value of the exponential function is $e^0 = 1$. At $t = 10$, however, the value of the exponential function is $e^{10} \cong (2.7183)^{10} \cong 22026$.

A more complete tabulation of the standard exponential function is given in Table III in Appendix A.

The exponential function can also be evaluated for *negative* values of t. Since

$$1 = e^0 = e^{(t-t)} = e^t e^{-t}$$

we find that e^{-t} is related to e^t by

$$e^{-t} = \frac{1}{e^t}$$

(4.35)

The functions e^t and e^{-t} are illustrated graphically in Figure 4.4 Note that e^t *increases* as *t increases,* whereas e^{-t} *decreases* as *t increases.*

Note also that the slope of the e^t *versus t* curve increases as *t* increases. Therefore, the *derivative* of e^t with respect to *t* is an *increasing* function of *t*. In fact, de^t/dt *increases exponentially* with increasing *t:*

$$\frac{d}{dt}(e^t) = e^t$$

(4.36)

FIGURE 4.4   Graphical representation of e^t and e^{-t} as functions of t.

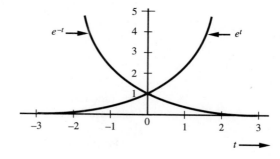

It can also be shown that

$$\frac{d}{dt}(e^{-t}) = -e^{-t}$$

(4.37)

Note that the slope of the e^{-t} versus t curve is *negative* (see Figure 4.4).

Equations (4.36) and (4.37) are indeed remarkable results. The derivative of an exponential function is directly proportional to the *same* exponential function!

The variable t in Equations (4.34) to (4.37) is a *dimensionless* independent variable. (It does *not* have the units of time!) The exponential functions that occur in physics are usually of the form

$$f(t) = e^{\alpha t}$$

where α is a constant and t is an independent variable *with specific units* (for example, *time*). The combination αt is *dimensionless;* therefore, α has the units of t^{-1}. If t is expressed in *seconds,* then α has units of $(seconds)^{-1}$. We often see the negative exponential function in the form

$$f(t) = e^{-t/\tau}$$

In this case, τ (tau) has the same units as t.

It is straightforward to calculate the derivative of $e^{\alpha t}$ with respect to t using the chain rule for differentiation (Equation 4.14). Let $g(f) = e^f$, and $f = \alpha t$, then

$$\frac{dg}{dt} = \frac{dg}{df}\frac{df}{dt}$$

$$= \frac{d}{df}(e^f) \times \frac{d}{dt}(\alpha t)$$

Since $de^f/df = e^f$, and $d(\alpha t)/dt = \alpha$, we find

$$\frac{dg}{dt} = e^f \times \alpha$$

Substituting $f = \alpha t$ and $g = e^f$, gives

$$\frac{d}{dt}(e^{\alpha t}) = \alpha e^{\alpha t}$$

(4.38)

In a similar manner, we can show that

$$\frac{d}{dt}(e^{-\alpha t}) = -\alpha e^{-\alpha t}$$

(4.39)

TABLE 4.4   Derivatives of Exponential Functions

$f(t)$	$\dfrac{df}{dt}$
e^t	e^t
$e^{\alpha t}$	$\alpha e^{\alpha t}$
e^{-t}	$-e^{-t}$
$e^{-\alpha t}$	$-\alpha e^{-\alpha t}$

The important derivative relations for exponential functions are summarized in Table 4.4 (see also Table 4.1).

The decay of a sample of radioactive material is an important application that involves the exponential function. If the initial intensity of the radioactivity is I_0 (at $t = 0$), it is observed that the intensity at time t is

$$I(t) = I_0 e^{-t/\tau} \tag{4.40}$$

where τ is a constant with dimensions of time and is called the *mean decay time.* The significance of τ is that after a time equal to τ, the activity has decreased to $1/e$ (approximately 36.8%) of its initial value. That is,

$$I(\tau) = I_0 \times \frac{1}{e} \cong 0.368\, I_0$$

This point is indicated on the curve of $I(t)$ *versus* t in Figure 4.5.

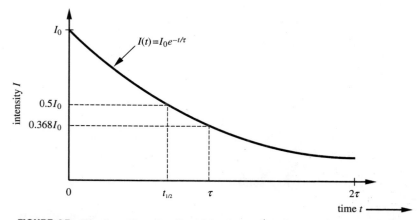

FIGURE 4.5   The intensity of radioactivity is $I_0 e^{-t/\tau}$ where τ is the mean decay time. The half-life of the sample is $t_{1/2} = 0.693\, \tau$.

The time required for the intensity to drop to one-half of its initial value is called the *half-life* of the decay. This time is denoted by $t_{1/2}$ and is related to the mean decay time τ by

$$t_{1/2} = 0.693\,\tau \qquad\qquad \textbf{(4.41)}$$

▶ **EXAMPLE 4.5.1**   A sample of the radioactive isotope ^{11}C is prepared in the laboratory. The half-life of ^{11}C is 20 min. At time $t = 0$, a radiation detector records a count rate of 1000 s^{-1} from the sample. (a) What is the mean decay time for ^{11}C? (b) What is the detector count rate at time $t = 52$ min?

The mean decay time is obtained from the half-life by using Equation (4.41).

$$\tau = \frac{t_{1/2}}{0.693} = \frac{20\ \text{min}}{0.693} = 28.8\ \text{min}$$

In order to answer part (b), we make use of Equation (4.40). Substituting $I_0 = 1000$ s^{-1}, $\tau = 28.8$ min, and $t = 52$ min, we find

$$I(52\ \text{min}) = (1000\ \text{s}^{-1}) \times e^{-52\ \text{min}/28.8\ \text{min}}$$

$$= (1000\ \text{s}^{-1}) \times e^{-1.8}$$

From Table III, in Appendix A, we find $e^{-1.8} = 0.165$. Therefore,

$$I(52\ \text{min}) = (1000\ \text{s}^{-1}) \times 0.165 = 165\ \text{s}^{-1}$$

4.6   LOGARITHMIC FUNCTIONS AND THEIR DERIVATIVES

The concept of a *logarithm* is a natural extension of the discussion of *exponents.* By definition,

> The logarithm to the base a of a number x *is equal to* the exponent y to which the base number a must be raised in order that $x = a^y$.   **(4.42)**

That is, if

$$x = a^y \qquad\qquad \textbf{(4.43)}$$

then

$$y = \log_a x \qquad\qquad \textbf{(4.44)}$$

In Equations (4.43) and (4.44), a is referred to as the *base number*. Equation (4.44) is the mathematical representation of the statement in Equation (4.42) and is to be read as

"y is equal to the logarithm to the base a of x"

Therefore, since $2^5 = 32$, we conclude that

$$5 = \log_2 32$$

Similarly, since $10^3 = 1000$, we find

$$3 = \log_{10} 1000$$

Since the logarithms of numbers are *exponents,* they have the same mathematical properties as exponents. For example, if

$$A = a^p \quad \text{and} \quad B = a^q$$

we find that

$$p = \log_a A \quad \text{and} \quad q = \log_a B$$

Therefore,

$$p + q = \log_a A + \log_a B$$

However, we also know (by generalizing Equation 1.1)

$$AB = a^p a^q = a^{p+q}$$

and so we find that

$$p + q = \log_a AB$$

The above expressions for $p + q$ can be combined to give the important identity,

$$\log_a AB = \log_a A + \log_a B \tag{4.45}$$

That is, the logarithm of the *product* of two numbers is equal to the *sum* of the logarithms of the individual numbers. The identity in Equation (4.45) forms the operational basis for multiplying two numbers, A and B. That is, we can *multiply* two numbers *by adding* their logarithms. In a similar manner (by generalizing Equation 1.4), it can be shown that

$$\log_a \frac{A}{B} = \log_a A - \log_a B$$

(4.46)

That is, the logarithm of the *ratio* of two numbers is equal to the *difference* of the logarithms of the individual numbers. Equation (4.46) forms the operational basis for *dividing* two numbers, *A* by *B*, namely, we *subtract* their logarithms. Finally, we leave it as an exercise for the student to verify the identity,

$$\log_a A^n = n \log_a A$$

(4.47)

That is, the logarithm of a number to a given exponent *n* is equal to *n* times the logarithm of the number.

Common Logarithms

Logarithms to the base 10 are called *common logarithms*. Rather than write $\log_{10} x$, it is customary to use the short-form notation,

$$\log x \quad \text{(common logarithm)}$$

to denote the common logarithm of *x*. A table of values of $\log x$ is given in Table IV of Appendix A for values of *x* ranging from 1.0 to 9.9, in steps of 0.1. For present purposes we have tabulated in Table 4.5 the values of $\log x$, for *x* ranging from 1 to 10, in steps of unity. Notice that the first and last values of $\log x$ in Table 4.5 follow from $10^0 = 1$ and $10^1 = 10$, respectively.

TABLE 4.5   Logarithms to the Base 10

x	log x
1	0.000
2	0.301
3	0.477
4	0.602
5	0.699
6	0.778
7	0.845
8	0.903
9	0.954
10	1.000

We now consider some simple numerical examples that make use of Table 4.5 and Equations (4.45) and (4.46).

▶ **EXAMPLE 4.6.1**

$$\log 2 + \log 4 = ?$$

From Equation (4.45) and Table 4.5, we find

$$\log 2 + \log 4 = \log (2 \cdot 4) = \log 8 = 0.903$$

Also, by direct computation,

$$\log 2 + \log 4 = 0.301 + 0.602 = 0.903$$

▶ **EXAMPLE 4.6.2**

$$\log 8 - \log 4 = ?$$

From Equation (4.46) and Table 4.5, we find

$$\log 8 - \log 4 = \log \frac{8}{4} = \log 2 = 0.301$$

Also, by direct computation,

$$\log 8 - \log 4 = 0.903 - 0.602 = 0.301$$

Note from Table 4.5 that $\log x$ is a *slowly varying* function of x. As x increases from 1 to 10, $\log x$ increases from 0 to 1. Moreover, since $\log 10^2 = 2 \log 10 = 2$, as x increases from 10 to 100, $\log x$ increases from 1 to 2. Similarly, as x increases from 100 to 1000, $\log x$ increases from 2 to 3, and so on. Since a large change in the value of x produces only a small change in the value of $\log x$, the function $\log x$ is said to be a *slowly varying* function of x. This property of $\log x$ has an important application in the graphical representation of functions that vary by several orders of magnitude.

Semilog and Log-Log Plots

In many problems encountered in physics, the dependent variable $f(t)$ (say) varies by several orders of magnitude when the independent variable t changes by a relatively small amount. An example that illustrates this point is the exponential decay of a sample of radioactive material discussed in Section 4.5 (see Equation 4.40 and Example 4.5.1). The intensity of the radioactivity at time t is

$$I(t) = I_0 e^{-t/\tau}$$

where I_0 is the initial intensity and τ is the mean decay time. At $t = 0$, the intensity is equal to I_0. By $t = 7\tau$, however, the intensity has decreased to the value (see Table III in Appendix A)

$$I(t = 7\tau) = I_0 e^{-7} = 0.0009 \, I_0 \cong 10^{-3} \, I_0$$

Therefore, in *seven* mean decay times the intensity has decreased by *three orders of magnitude* relative to its initial value. Since $I(t)$ changes by such a large amount, how do we provide a meaningful graphical representation of $I(t)$ *versus* t during the time interval from $t = 0$ to $t = 7\tau$? It is clear from Figure 4.5 that a graphical representation in which both the vertical and horizontal scales are *linear* or *uniform** (that is, *directly proportional* to the values of I and t, respectively) provides very poor resolution at small intensities ($I \lesssim 10^{-1} \, I_0$, say). To alleviate this difficulty it is useful to make a *semilogarithmic* (or *semilog*) plot of $I(t)$ *versus* t (see Example 4.6.3). In this case, the horizontal scale (t) is linear. However, the *vertical scale is proportional to log I*, rather than I. Because log I is a *slowly varying* function of I, the vertical size of the graph remains compact even though I varies over several orders of magnitude. Furthermore, the semilog plot provides very good resolution of I at small values.

▶ **EXAMPLE 4.6.3** Consider the decay of the radioactive isotope ^{11}C discussed in Example 4.5.1. The initial count rate is $I_0 = 10^3 \text{ s}^{-1}$, and the mean decay time of the sample is $\tau = 28.8$ minutes. Make a semilog plot of the count rate $I(t)$ *versus* time t for the time interval from $t = 0$ to $t = 7\tau$.

For this example, the count rate (in s^{-1}) at time t is (see Equation 4.40)

$$I(t) = I_0 e^{-t/\tau}$$
$$= 10^3 \, e^{-t/\tau} \qquad \text{(1)}$$

where $\tau = 28.8$ min. The semilog plot of $I(t)$ *versus* t is indicated by the solid straight line in the accompanying figure. The horizontal and vertical lines are *grid lines* (or *mesh lines*) characteristic of semilog graph paper.

---

*A *linear* or *uniform* scale is one in which the grid lines are equally spaced (see, for example, the horizontal scale in the figure accompanying Example 4.6.3).

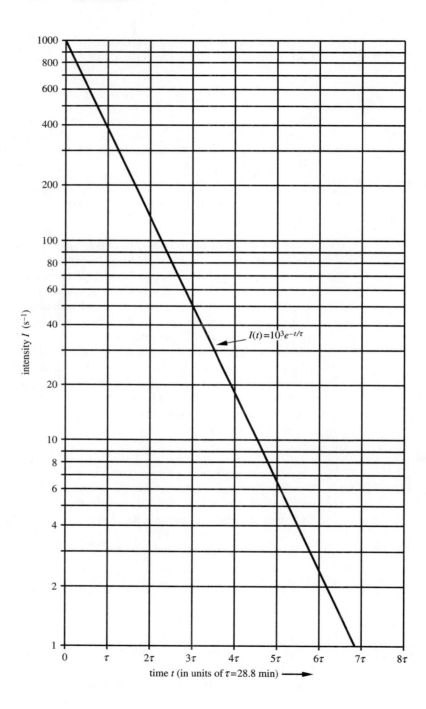

The following important features of the graph should be noted:

(a) The scale along the horizontal axis (the *time* axis) is *linear*. (The spacing between the vertical grid lines is equal). Moreover, t is plotted in units of the mean decay time τ.

(b) The intensity I, plotted along the vertical axis, decreases by *three* orders of magnitude from $I(t = 0) = 1000 \text{ s}^{-1}$ to $I(t = 7\tau) \cong 1 \text{ s}^{-1}$. The scale along the vertical axis is *not* linear. That is, the spacings between the horizontal grid lines are *not* equal. Rather, the *vertical scale is proportional to log I*, and the spacing *between* any two grid lines is proportional to the *difference* of their logarithms. Let us denote by L the distance between the grid lines representing $I = 1000$ and $I = 100$ (*or* between $I = 100$ and $I = 10$, *or* between $I = 10$ and $I = 1$). Making use of Equation (4.45) and Table 4.5, we first note that

$$(\log 1000 - \log 100) \times L = (\log 10 + \cancel{\log 100} - \cancel{\log 100}) \times L$$

$$= L \qquad \text{(2)}$$

That is, the distance between the grid lines representing $I = 1000$ and $I = 100$ is proportional to $\log 1000 - \log 100$, with constant of proportionality equal to L. Similarly, the distance between the grid lines representing $I = 200$ and $I = 100$ is

$$(\log 200 - \log 100) \times L = (\log 2 + \cancel{\log 100} - \cancel{\log 100}) \times L$$

$$= 0.301 \, L \qquad \text{(3)}$$

Moreover, the distance between the grid lines representing $I = 600$ and $I = 200$ is

$$(\log 600 - \log 200) \times L = (\log 6 + \cancel{\log 100} - \log 2 - \cancel{\log 100}) \times L$$

$$= (0.778 - 0.301) \times L$$

$$= 0.477 \, L \qquad \text{(4)}$$

and so on. The student should *measure* the distance between the appropriate grid lines in the accompanying figure, and verify directly that (3) and (4) are correct.

It should also be noted that the vertical grid in semilog plots is *cyclic*. (In the accompanying figure there are *three* cycles). The spacing between corresponding pairs of grid lines in each cycle is identical. For example, the spacing between the grid lines representing $I = 6$ and $I = 2$ is identical to the spacing between the $I = 600$ and $I = 200$ grid lines. This follows because

$$(\log 6 - \log 2) \times L = (0.778 - 0.301) \times L$$
$$= 0.477\,L \tag{5}$$

which is identical to (4).

(c)   The *semilog* plot of the exponential function $I_0 e^{-t/\tau}$ *versus t* gives a *straight line,* rather than a *curved line* as in Figure 4.5. The reason for this is the following. The vertical scale is proportional to $\log I$ where

$$\log I = \log (I_0 e^{-t/\tau})$$
$$= \log I_0 + \log (e^{-t/\tau})$$
$$= \log I_0 + \log \left(\frac{1}{e^{t/\tau}}\right) \tag{6}$$

Making use of Equations (4.46) and (4.47) gives

$$\log I = \log I_0 + \log 1 - \log (e^{t/\tau})$$
$$= \log I_0 + 0 - \frac{t}{\tau} \log e \tag{7}$$

To four-figure accuracy, the logarithm of *e to the base* 10 is

$$\log e = 0.4343 \tag{8}$$

It follows from (7) and (8) that

$$\log I = \log I_0 - \frac{0.4343}{\tau} t \tag{9}$$

Therefore, the graph of $\log I$ versus t is a *straight line* with

$$\text{slope} = -\frac{0.4343}{\tau} \tag{10}$$

The student should compute the slope of the straight line in the accompanying figure and show that (10) is indeed satisfied.

Since the values of I_0 and τ have not been specified in obtaining (9) and (10), the results are quite general. In practice, (10) can be used to compute the mean decay time τ for a radioactive sample. The count rate I is measured experimentally and plotted as a function of time t on semilog graph paper. The slope of the resulting straight line is then measured, and (10) is used to compute τ.

In general, a plot of $f(t)$ *versus* t in which the vertical scale (f) is *logarithmic* and the horizontal scale (t) is *linear* is called a *semilogarithmic* (or *semilog*) plot of $f(t)$ *versus* t. It is important to note that the semilog plot of any exponential function,

$$f(t) = a^t$$

where a is a constant, gives a *straight line*. Making use of Equation (4.47) it follows that

$$\log f = \log (a^t)$$
$$= t \log a$$

Therefore the graph of $\log f$ versus t is a straight line with slope equal to $\log a$.

In circumstances where *both* the dependent variable $f(t)$ and the independent variable t vary by several orders of magnitude it is useful to make a *log-log* (or *logarithmic*) plot of $f(t)$ *versus* t. In this case, both the vertical scale (f) and the horizontal scale (t) are *logarithmic*. The basic concepts involved in making log-log plots are similar to those discussed in Example 4.6.3. Therefore, for present purposes, it is adequate to illustrate with a brief example.

▶ **EXAMPLE 4.6.4** Newton's law of universal gravitation states that the gravitational force between two bodies with masses m_1 and m_2 that are separated by a distance r is

$$F(r) = \frac{Gm_1m_2}{r^2} \tag{1}$$

where G = gravitational constant = 6.673×10^{-11} N·m²/kg. Consider a spacecraft with mass

$$m_1 = 10^4 \text{ kg} \tag{2}$$

located in the gravitational field of the Earth. The mass and radius of the Earth are

$$m_2 = 5.98 \times 10^{24} \text{ kg} \tag{3}$$

and

$$R_E = 6.38 \times 10^6 \text{ m} \tag{4}$$

respectively (see Table VI in Appendix A). Make a log-log plot of F *versus* r for distances ranging from $r = R_E$ to $r = 100\ R_E$.

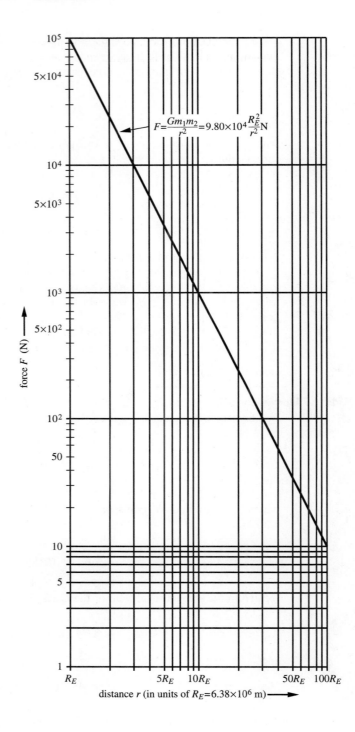

$$F = \frac{Gm_1m_2}{r^2} = 9.80 \times 10^4 \frac{R_E^2}{r^2} \text{N}$$

force F (N) ⟶

distance r (in units of $R_E = 6.38 \times 10^6$ m) ⟶

For this example, the independent variable r varies by *two* orders of magnitude. Combining (1)–(4), it is straightforward to show that

$$F(r) = \frac{Gm_1m_2}{R_E^2}\frac{R_E^2}{r^2}$$

$$= 9.80 \times 10^4 \frac{R_E^2}{r^2} \text{ N} \tag{5}$$

Therefore

$$F(r = R_E) = 9.80 \times 10^4 \text{ N} \tag{6}$$

and

$$F(r = 100 \ R_E) = 9.80 \text{ N} \tag{7}$$

Note that the dependent variable $F(r)$ varies by *four* orders of magnitude for the range of r under consideration,.

The log-log plot of $F(r)$ *versus* r is indicated by the solid straight line in the accompanying figure. The horizontal and vertical lines are *grid lines* (or *mesh lines*) · characteristic of log-log graph paper. (Not all of the grid lines are shown in the figure.)

Note that *both* the horizontal and vertical scales are *logarithmic*. That is, the horizontal scale is proportional to log r and the vertical scale is proportional to log F. This is in contrast to a semilog plot where only the vertical scale is logarithmic (see Example 4.6.3).

Note also that the log-log plot of Gm_1m_2/r^2 *versus* r gives a *straight line* (rather than a *curved line*). The reason for this is the following. The vertical scale is proportional to log F, and the horizontal scale is proportional to log r. From (1)

$$\log F = \log\left(\frac{Gm_1m_2}{r^2}\right)$$

Making use of Equations (4.46) and (4.47), we find

$$\log F = \log (Gm_1m_2) - \log r^2$$

$$= \log (Gm_1m_2) - 2\log r \tag{8}$$

Therefore, the graph of log F *versus* log r is a straight line with

$$\text{slope} = -2 \tag{9}$$

The student should compute the slope of the straight line in the accompanying figure and verify that (9) is indeed satisfied. Since the values of m_1 and m_2 have not been specified in obtaining (8) and (9), the results are quite general.

It is important to note that the log-log plot of any algebraic function,

$$f(t) = t^a$$

where a is a constant, gives a *straight line*. From Equation (4.47) it follows that

$$\log f = a \log t$$

Therefore the graph of $\log f$ *versus* $\log t$ is a straight line with slope equal to a.

Natural Logarithms

Although logarithms to the base 10 are important for understanding the principle of semilog and log-log plots and for general calculation, the base encountered most frequently in scientific work is *not* 10. Rather it is the same constant e used to define the standard exponential function in Equation (4.34). Logarithms to the base e are called *natural* logarithms or *Napierian* logarithms. Rather than write $\log_e x$, it is customary to use the short-form notation

$$\ln x \text{ (natural logarithm)}$$

to denote the natural logarithm of x.

Consider the equation

$$y = \ln x \qquad \text{(4.48)}$$

From the basic definition of logarithms in Equation (4.42), Equation (4.48) means that y is the power to which e must be raised to give x. That is,

$$x = e^y \qquad \text{(4.49)}$$

The function $y = \ln x$ is illustrated graphically in Figure 4.6. At $x = 0$, $\ln x$ approaches $-\infty$, and as x approaches $+\infty$, $\ln x$ approaches $+\infty$. The value of $\ln x$ is equal to zero at $x = 1$, and it is equal to 1 at $x = e$. The function $\ln x$ is tabulated in Table V of Appendix A.

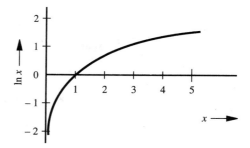

FIGURE 4.6   Graph of ln x *versus* x.

Figure 4.6 shows that the slope of the ln x *versus* x curve decreases as x increases. The derivative of ln x with respect to x is

$$\frac{d}{dx}(\ln x) = \frac{1}{x} \tag{4.50}$$

which is a decreasing function of x. Moreover,

$$\frac{d}{dx}(\ln \alpha x) = \frac{1}{x} \tag{4.51}$$

where α is any constant.

In concluding this section we summarize several important properties of logarithmic functions:

$$\ln \frac{x}{y} = \ln x - \ln y$$

$$\ln xy = \ln x + \ln y$$

$$\ln a^x = x \ln a$$

$$\ln e^x = x$$

$$\ln e = 1 \tag{4.52}$$

EXERCISES   *Evaluate the following: (Hint: Use Table* **III**, *Appendix A.)*

1. e^{-3t}, for $t = 2$.   (Ans. 338)

2. $\frac{d}{dt}(e^{2t})$, for $t = 1.5$.   (Ans. 249)

3. e^{6t}, for $t = 1$.   (Ans. 175)

4. $\frac{d}{dt}(e^{-5t})$, for $t = 0.5$.   (Ans. 384)

5. Verify the identity in Equation (4.46).

6. Verify the identity in Equation (4.47).

7. $\log 2^n = $ _____   (Ans. 156)

8. $\log 7 - \log 3.5 = $ _____   (Ans. 237)

9. $\log 27 = 9 \log$ _____   (Ans. 280)

10. $\log 25 = $ _____   (Ans. 236)

11. $\log 5 + \log 2 = $ _____   (Ans. 36)

12. Make a semilog plot of $2^{t/\tau}$ *versus* time t for the interval from $t = 0$ to $t = 10\tau$, where $\tau = $ constant = 10 min.

13. Make a log-log plot of $(t/\tau)^2$ *versus* time t for the interval from $t = \tau$ to $t = 100\tau$, where $\tau = $ constant = 10 min.

14. Verify the identity in Equation (4.51).

15. Verify that $\ln e = 1$.

16. Evaluate $\dfrac{d^2}{dx^2} \ln \alpha x$ for arbitrary x.   (Ans. 332)

17. Evaluate $\dfrac{d}{dx} \ln (10^6 x)$ for $x = e$.   (Ans. 61)

CHAPTER

5

Vectors

5.1 PROPERTIES OF VECTORS

In discussions of physical situations we find use for a number of quantities that are completely specified in terms of *magnitude* only—for example, mass, time, and temperature. Such quantities are called *scalars*. Some of the common scalars are listed in Table 5.1.

Certain physical quantities are not completely specified by magnitude alone. To say that the speed of an automobile is 60 km/h omits an important part of the motion, namely, the *direction* in which the automobile is moving. Quantities that require both magnitude and direction for a complete description are called *vectors*. It is customary to represent a vector by "bold-face" type, such as **A**. Some of the common vectors are listed in Table 5.1.

As shown in Figure 5.1, a vector **A** can be represented pictorially by a *directed line segment* from some origin O (the *foot* of the vector) to a point P (the *head* of

TABLE 5.1 Some Common Scalars and Vectors

Scalars	Vectors
Distance, x, s	Displacement, **x, s**
Speed, v	Velocity, **v**
Mass, m	Momentum, $\mathbf{p} = m\mathbf{v}$
Time, t	Force, **F**
Work, W	Acceleration, **a**
Volume, V	Electric field, **E**
Temperature, T	Magnetic field, **B**

FIGURE 5.1   Pictorial representation of a vector **A**.

the vector). The vector **A** is sometimes denoted by **OP**. The direction of the straight line segment, as inferred by the direction of the arrowhead in Figure 5.1, corresponds to the *direction* of the vector **A**. The length of the straight line segment is proportional to the *magnitude* of **A**, in some agreed units. It is customary to denote the magnitude of **A** by $|\mathbf{A}|$, or simply by A. That is,

$$\text{Magnitude of vector } \mathbf{A} = |\mathbf{A}| \text{ or } A \qquad (5.1)$$

The magnitude of a *vector* **A** is the *scalar A*. (The magnitude of *velocity* is *speed*.)

A vector **A** can be multiplied directly by a scalar b. The resulting quantity is also a vector, which we denote by **A′**, where

$$\mathbf{A}' = b\mathbf{A} \qquad (5.2)$$

The direction of **A′** is either *parallel* or *antiparallel* to **A**, depending on whether b is a positive number or a negative number, respectively. For example, if $b = 2.5$, then

$$\mathbf{A}' = 2.5\ \mathbf{A}$$

On the other hand, if $b = -1$, then

$$\mathbf{A}' = -\mathbf{A}$$

The vectors 2.5 **A**, −**A**, together with the vector **A**, are shown in Figure 5.2.

The vector 2.5 **A** is parallel to the vector **A**; that is, the new vector points in the *same direction* as **A**. The vector −**A**, however, is antiparallel to **A**, since it points in the direction *opposite* to **A**. Furthermore, keeping in mind that the magnitude of a vector is always a *positive* number, or zero (if **A** is a *null* vector, **A** = 0), we find that the magnitude of the vectors 2.5 **A** and −**A** are, respectively, 2.5 A and A.

FIGURE 5.2   The vectors **A**, 2.5**A**, and −**A**. The vector −**A** is antiparallel to the other two vectors.

Referring to Equation (5.2), we find in the general case that the magnitude of the vector $\mathbf{A}'$ is equal to $|b|A$, where $|b|$ is the *magnitude* of the number b; that is,

$$A' = |b|A \qquad \text{(5.3)}$$

EXERCISES

*The vector v is a velocity of **30 km/h** directed 45° north of due east. What are the magnitudes and directions of the following vectors?*

1. 3 **v**   (Ans. 303)   **3.** 1.5 **v**   (Ans. 165)
2. −3 **v**   (Ans. 301)   **4.** −2 **v**   (Ans. 285)

*The vector **F** is a force of **10³ N** directed vertically downward on a horizontal floor. What are the magnitudes and directions of the following vectors?*

5. 100 **F**   (Ans. 176)   **7.** −6 **F**   (Ans. 134)
6. 12 **F**   (Ans. 84)   **8.** −30 **F**   (Ans. 172)

5.2 ADDITION AND SUBTRACTION OF VECTORS

In a flat portion of Kansas, a motorist drives 10 kilometers due east from point O to point P. His *displacement* during this portion of the journey is represented by the vector $\mathbf{A} = \mathbf{OP}$ in Figure 5.3. At P the motorist turns left and then drives 20 kilometers due north to point P', his final destination. This displacement is represented by the vector $\mathbf{B} = \mathbf{PP}'$ in Figure 5.3. It is clear that the trip from O to P, followed by the trip from P to P', has the same end result as a trip *directly* from O to P'. As shown in the figure, the *net* displacement for the entire trip can be represented by the vector $\mathbf{R} = \mathbf{OP}'$.

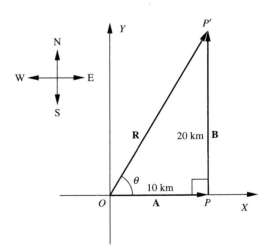

FIGURE 5.3   A journey in Kansas from O to P to P'.

The vector displacement **R** is called the *resultant* of the two vector displacements, **A** and **B**. We can express **R** in equation form as

$$\mathbf{R} = \mathbf{A} + \mathbf{B} \tag{5.4}$$

Equation (5.4) is a *vector equation,* and states that the resultant vector displacement **R** is equal to the sum of the two vector displacements, **A** and **B**. We note from Figure 5.3 that the vectors **A**, **B**, and **R** together form the three sides of the triangle *OPP'*.

The magnitude and direction of the resultant vector **R** can be determined in the following manner. Referring to Figure 5.3, and making use of the fact that *OPP'* is a *right* triangle with angle $\angle OPP' = 90°$, we find, from the Pythagorean theorem,

$$R = \text{magnitude of resultant vector } \mathbf{R} \text{ in Figure 5.3}$$
$$= \sqrt{(10)^2 + (20)^2}\,\text{km} = \sqrt{500}\,\text{km} = 22.36\,\text{km}$$

Next, we let θ represent the angle $\angle P'OP$. From the definition of tan θ,

$$\tan\theta = \frac{20\,\text{km}}{10\,\text{km}} = 2$$

We can estimate $\theta = 63.4°$ from the values in Table I of Appendix A, or we can use a calculator.

Therefore, we have found that the resultant vector **R** is equal to a displacement of 22.36 kilometers directed 63.4° north of due east. We emphasize that to determine completely the vector **R**, *both* its magnitude and direction must be specified.

The above example illustrates a further important property of vector addition. The vector equation, $\mathbf{R} = \mathbf{A} + \mathbf{B}$, does *not* imply that the magnitude of the resultant vector **R** is equal to the sum of the magnitudes of the vectors **A** and **B**. In Figure 5.3 we see that $A + B = 10\,\text{km} + 20\,\text{km} = 30\,\text{km}$, whereas $R = 22.36$ km. Therefore, even though $\mathbf{R} = \mathbf{A} + \mathbf{B}$ is true, $R = A + B$, is *not* true. This is, in fact, a general property of vector addition, and is not restricted to the specific example illustrated in Figure 5.3.

The example discussed above also serves to demonstrate the general method for finding the resultant of any two vectors **A** and **B** which are *orthogonal,* that is, *perpendicular* to one another.

We now develop a procedure for determining the resultant **R** of any two vectors **A** and **B** with *arbitrary* orientations. The procedure is known as the *method of triangles.* Consider the vectors **A** and **B** represented pictorially in Figures 5.4(*a*) and (*b*). The vectors **A** and **B** can signify displacements, velocities, momenta, forces,

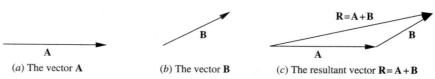

(*a*) The vector **A**   (*b*) The vector **B**   (*c*) The resultant vector **R** = **A** + **B**

FIGURE 5.4   Constructing the triangle of vectors to determine **R** = **A** + **B**.

FIGURE 5.5 Two possible triangles for constructing the resultant vector,
$\mathbf{R} = \mathbf{A} + \mathbf{B}$.

or any other type of vector. **A** and **B** must, however, represent the same type of quantity if the sum **A** + **B** is to be physically meaningful.

As shown in Figure 5.4(*c*), we determine the resultant vector $\mathbf{R} = \mathbf{A} + \mathbf{B}$ by constructing a triangle in which **A** and **B** are adjacent sides and then connect head to foot. The resultant vector **R** is equal to the third side of the triangle. That is, the magnitude of **R** corresponds to the length of the third side, and the direction of **R** is that indicated in Figure 5.4(*c*).

It is important to note that the order in which **A** and **B** are connected does not affect the resultant **R**. The two possibilities are illustrated in Figure 5.5. In Figure 5.5, the resultant vector **R** has the same magnitude and direction for both triangles. Therefore, in practice we can construct **R** by either method indicated in Figure 5.5. This important result is a reflection of the fact that vector addition obeys the *commutative law;* that is,

$$\mathbf{A} + \mathbf{B} = \mathbf{B} + \mathbf{A} \tag{5.5}$$

The procedure illustrated in Figures 5.4 and 5.5 can be extended to determine the resultant **R** of any number of vectors **A**, **B**, **C**, ... For example, to determine the resultant $\mathbf{R} = \mathbf{A} + \mathbf{B} + \mathbf{C} + \mathbf{D} + \mathbf{E}$ for the five vectors **A**, **B**, **C**, **D**, and **E**, we construct the hexagon (6-sided polygon) shown in Figure 5.6. In this figure the vectors **A**, **B**, **C**, **D**, and **E** are connected together, head to foot, and form five adjacent sides of a hexagon. The resultant vector **R** is then equal to the sixth side; that is, the magnitude of **R** corresponds to the length of the sixth side, and the direction of **R** is that indicated in Figure 5.6. As before, the order in which the vectors **A**, **B**, **C**, **D**, and **E** are connected does not affect the magnitude or direction

FIGURE 5.6 A hexagon of vectors for constructing the resultant vector, $\mathbf{R} = \mathbf{A} + \mathbf{B} + \mathbf{C} + \mathbf{D} + \mathbf{E}$.

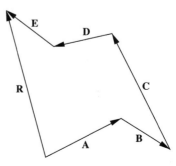

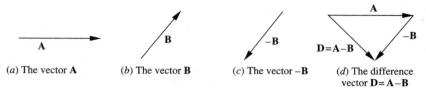

FIGURE 5.7   Constructing the triangle of vectors to determine $\mathbf{D} = \mathbf{A} - \mathbf{B}$.

of $\mathbf{R}$. (Verify this statement by trying several different orderings for adding the vectors. Can you make use of the fact that $\mathbf{E} = -\mathbf{B}$?)

The procedure for *subtracting* a vector $\mathbf{B}$ from a vector $\mathbf{A}$ is straightforward. We first reverse the direction of $\mathbf{B}$, that is, we form the vector $-\mathbf{B}$, and then *add* it to $\mathbf{A}$. The difference vector $\mathbf{D}$ is denoted by

$$\mathbf{D} = \mathbf{A} - \mathbf{B} \qquad (5.6)$$

Since we can write $-\mathbf{B} = +(-\mathbf{B})$, Equation (5.6) can also be expressed in the form

$$\mathbf{D} = \mathbf{A} + (-\mathbf{B}) \qquad (5.7)$$

That is, $\mathbf{D}$ is equal to the *sum* of the vectors $\mathbf{A}$ and $-\mathbf{B}$. As illustrated in Figure 5.7, we determine the difference vector, $\mathbf{D} = \mathbf{A} - \mathbf{B}$, by constructing a vector triangle in which $\mathbf{A}$ and $-\mathbf{B}$ form adjacent sides. The vector $\mathbf{D}$ is then equal to the third side of the triangle. That is, the magnitude of $\mathbf{D}$ corresponds to the length of the third side, and the direction of $\mathbf{D}$ is that indicated in Figure 5.7. We reiterate that the difference vector, $\mathbf{D} = \mathbf{A} - \mathbf{B}$, is equal to the resultant of the vectors $\mathbf{A}$ and $-\mathbf{B}$.

EXERCISES

The vector x *is a displacement of* **45 m** *directed* **45° north of due west. What are the magnitudes and directions of the following vectors?**

1. x + 2 x   (Ans. 288)
2. x − 2 x   (Ans. 212)
3. x + 2 x + 3 x   (Ans. 178)
4. x + 2 x − 3 x   (Ans. 25)
5. x + 2 x + 3 x + 4 x   (Ans. 39)

6. As shown in the figure, a pilot is attempting to fly his airplane due north with a velocity **v** equal to 600 km/h. There is a crosswind with $\mathbf{v}_{wind}$ equal to 200 km/h from the west. What is the magnitude and direction of the resultant velocity $\mathbf{v}_R$ of the airplane? (Ans. 28)

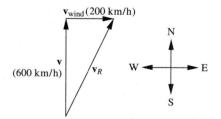

5.3   VECTOR COMPONENTS

By using the basic rules of trigonometry, any vector **A** can be *decomposed* into two vectors, the sum of which yields the original vector. This process is especially useful when the two vectors, which are called *component* vectors, are at right angles to one another. In this case, the two component vectors are referred to as *rectangular* component vectors. The general procedure for *resolving* a vector **A** into two rectangular components, A_x and A_y, is illustrated in Figure 5.8.

In Figure 5.8(*a*), we let θ be the angle between the positive *X*-axis and the vector **A**. In Figure 5.8(*b*), from the head of vector **A** (that is, the point *P*) we drop a perpendicular *PQ* to the *X*-axis. The angle $\angle OQP$ is then equal to 90°. It is evident from Figure 5.8(*b*) that *OPQ* forms a triangle of vectors consisting of A_x, A_y, and **A**. Furthermore,

$$\mathbf{A} = \mathbf{A}_x + \mathbf{A}_y \tag{5.8}$$

That is, the vector sum of A_x and A_y is equal to the vector **A**. The original vector **A** in Figure 5.8(*a*) can therefore be *removed* and *replaced* by the two rectangular components, A_x and A_y. We emphasize that the sum of two component vectors, A_x and A_y in Figure 5.8(*c*), is *totally equivalent* to the original vector **A** in Figure 5.8(*a*).

In Figure 5.8(*b*) the triangle *OQP* is a right triangle with $\angle OQP = 90°$, and $\angle POQ = \theta$. From the definition of cos θ, we find

$$\cos \theta = \frac{OQ}{OP} = \frac{A_x}{A} \tag{5.9}$$

Similarly, from the definition of sin θ, we find

$$\sin \theta = \frac{PQ}{OP} = \frac{A_y}{A} \tag{5.10}$$

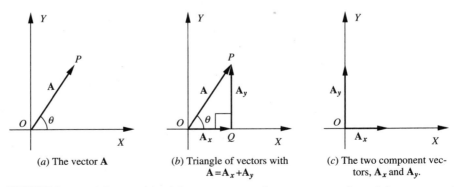

(*a*) The vector **A**

(*b*) Triangle of vectors with $\mathbf{A} = \mathbf{A}_x + \mathbf{A}_y$

(*c*) The two component vectors, $\mathbf{A}_x$ and $\mathbf{A}_y$.

FIGURE 5.8   Resolving a vector **A** into two rectangular components, A_x and A_y:
$$\mathbf{A} = \mathbf{A}_x + \mathbf{A}_y.$$

Therefore, if the magnitude A of the original vector **A** and the angle θ which **A** makes with the positive X-axis are specified, the rectangular components, A_x and A_y, can be determined from the equations

$$A_x = A \cos \theta$$

$$A_y = A \sin \theta$$

(5.11)

▶ **EXAMPLE 5.3.1**   If **A** represents a force vector of 50 N directed at an angle of 60° with respect to the positive X-axis, what are the values of A_x and A_y?

Evidently,

$$A = 50 \text{ N, and } \theta = 60°$$

From Equation (5.11) we find

$$A_x = (50 \text{ N}) \cos 60° = (50 \text{ N}) \cdot (0.5) = 25 \text{ N}$$

and

$$A_y = (50 \text{ N}) \sin 60° = (50 \text{ N}) \cdot (0.866) = 43.3 \text{ N}$$

Thus, the original force vector is equivalent to a force of 25 N in the positive X-direction, combined with a force of 43.3 N in the positive Y-direction.

▶ **EXAMPLE 5.3.2**   Find the values of A_x and A_y in the situation illustrated in the accompanying figure.

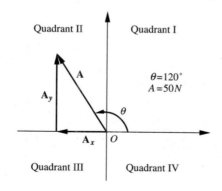

Referring to the figure above and to Equation (5.11), we find

$$A_y = (50 \text{ N}) \sin 120°$$

(1)

However, from Equations (2.38), we can write

$$\sin 120° = \sin(180° - 120°) = \sin 60° = 0.866 \tag{2}$$

Therefore,

$$A_y = (50 \text{ N}) \cdot (0.866) = 43.3 \text{ N} \tag{3}$$

Furthermore, referring again to the figure and to Equation (5.11), we find

$$A_x = (50 \text{ N}) \cos 120°$$

But from Equation (2.38), we note that the cosine function is negative in Quadrant II, and that

$$\cos 120° = -\cos(180° - 120°) = -\cos 60° = -0.5$$

Therefore,

$$A_x = (50 \text{ N}) \cdot (-0.5) = -25 \text{ N} \tag{4}$$

The fact that the value of A_x given in (4) is a *negative* number is simply a reflection of the fact that the component vector $\mathbf{A}_x$ is in the *negative* X-direction. Note that the value of A_y given in (3) is a *positive* number, since the component vector $\mathbf{A}_y$ is in the *positive* Y-direction.

The complete determination of the vector $\mathbf{A}$ (magnitude and direction) can be obtained from the rectangular components, A_x and A_y. First, we note that

$$A = \sqrt{A_x^2 + A_y^2} \tag{5.12}$$

and we use Equations (5.11) to write

$$\frac{A_y}{A_x} = \frac{A \sin \theta}{A \cos \theta} = \tan \theta$$

That is,

$$\tan \theta = \frac{A_y}{A_x} \tag{5.13a}$$

or

$$\theta = \tan^{-1} (A_y/A_x) \tag{5.13b}$$

which completes the specification of $\mathbf{A}$ in terms of A_x and A_y.

▶ **EXAMPLE 5.3.3**   For $A_x = 3$ cm and $A_y = 4$ cm, what are the values of A and θ?
From Equation (5.12),

$$A = \sqrt{3^2 + 4^2}\ \text{cm} = \sqrt{25}\ \text{cm} = 5\ \text{cm}$$

and from Equation (5.13),

$$\tan \theta = \frac{4\ \text{cm}}{3\ \text{cm}} = 1.333$$

so that

$$\theta = 53.1°$$

Thus, the vector **A** represents a displacement of 5 cm directed at an angle of 53.1° with respect to the positive X-axis.

EXERCISES   *Determine the quantities A, θ, A_x, and A_y for the following vectors **A**.*

1. **A** represents a force of 10 N along the positive X-axis.
(Ans. 151)
2. **A** represents a displacement of 20 m along the positive Y-axis.
(Ans. 122)
3. **A** represents a momentum of 10 kg · m/s directed at an angle of 210° from the X-axis.
(Ans. 371)
4. **A** represents a velocity of 320 m/s directed at an angle of 96° from the X-axis.
(Ans. 217)

*What is the magnitude and what is the direction of the force **F** in the following cases?*

5. The rectangular components of **F** are $F_x = 10$ N and $F_y = 20$ N.
(Ans. 136)
6. The rectangular components of **F** are $F_x = -20$ N and $F_y = 10$ N.
(Ans. 117)

5.4   ADDITION OF VECTORS BY COMPONENTS

The resolution of vectors into rectangular components, as discussed in Section 5.3, provides us with a powerful tool for the addition of vectors. Consider the two vectors, **A** and **B**, which are to be added together, as shown in Figure 5.9, to form the resultant vector **R** = **A** + **B**. The diagram, because it shows in detail the method of constructing the components, appears cluttered. But the process of vector addition by components is, in fact, quite simple. The x-component of the resultant vector is just the sum of the x-components of the two vectors being added; and similarly for the y-components. That is,

$$R_x = A_x + B_x$$
$$R_y = A_y + B_y$$

(5.14)

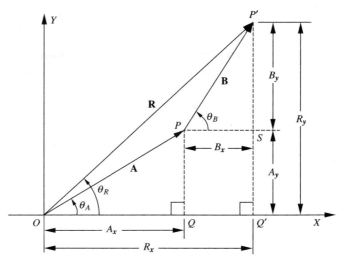

FIGURE 5.9   Resolving the vectors **A**, **B**, and **R** = **A** + **B**, into rectangular components.

Furthermore, the magnitude R of the resultant vector **R** is obtained from the Pythagorean theorem:

$$R = \sqrt{R_x^2 + R_y^2}$$
$$= \sqrt{(A_x + B_x)^2 + (A_y + B_y)^2} \qquad \text{(5.15)}$$

Finally, we can obtain the angle θ_R that defines the direction of **R** from

$$\tan \theta_R = \frac{R_y}{R_x} = \frac{A_y + B_y}{A_x + B_x} \qquad \text{(5.16)}$$

If, in some problem, the quantities given are A, θ_A, B, and θ_B, then we write

$$A_x = A \cos \theta_A$$
$$A_y = A \sin \theta_A$$
$$B_x = B \cos \theta_B$$
$$B_y = B \sin \theta_B$$

and proceed according to the above discussion to determine R and θ_R.

This procedure can be extended to the sum of three or more vectors in an obvious way.

▶ **EXAMPLE 5.4.1**   Suppose that a driver travels 10 kilometers due east and then turns left through an angle of 60° and travels 20 additional kilometers. What vector, originating at the driver's starting point, specifies the final position?

Referring to Figure 5.9, we have

$$A = 10 \text{ km}; \quad \theta_A = 0°$$
$$B = 20 \text{ km}; \quad \theta_B = 60°$$

Therefore,

$$A_x = A \cos \theta_A = (10 \text{ km}) \cos 0° = 10 \text{ km}$$
$$A_y = A \sin \theta_A = (10 \text{ km}) \sin 0° = 0$$

Similarly,

$$B_x = B \cos \theta_B = (20 \text{ km}) \cos 60° = (20 \text{ km}) \cdot 0.5 = 10 \text{ km}$$
$$B_y = B \sin \theta_B = (20 \text{ km}) \sin 60° = (20 \text{ km}) \cdot 0.866 = 17.32 \text{ km}$$

Substituting into Equation (5.15), we find that the magnitude of the resultant vector **R** is

$$R = \sqrt{(10 + 10)^2 + (0 + 17.32)^2} \text{ km} = \sqrt{(20)^2 + (17.32)^2} \text{ km}$$
$$= \sqrt{700} \text{ km} = 26.5 \text{ km}$$

Furthermore, substituting into Equation (5.16), we find

$$\tan \theta_R = \frac{0 + 17.32}{10 + 10} = \frac{17.32}{20} = 0.866$$

so that

$$\theta_R = 40.9°$$

EXERCISES   *Find R_x, R_y, R, and θ_R in the following cases:*

1. $A_x = 2$ m, $A_y = 3$ m; $B_x = 3$ m, $B_y = 2$ m.
(Ans. 281)

2. $A_x = 53$ m/s, $A_y = 100$ m/s; $B_x = -53$ m/s, $B_y = 230$ m/s.
(Ans. 352)

Find A_x, A_y, B_x, B_y, R_x, R_y, R and θ_R in the following cases:

3. $A = 100$ N; $\theta_A = 30°$; $B = 75$ N, $\theta_B = 60°$.

(Ans. 360)

4. $A = 25$ m, $\theta_A = 45°$; $B = 25$ m, $\theta_B = 90°$.

(Ans. 257)

5.5   UNIT VECTORS

The procedures for dealing with vector components can be simplified by introducing the idea of *unit vectors*. A general vector **A** can have any direction—it can lie along an axis or it can make an arbitrary angle with an axis. A unit vector, on the other hand, always lies exactly along an axis, and one such unit vector is assigned to each axis of the coordinate system (Fig. 5.10). For two-dimensional problems, there are two unit vectors (for the *X*- and *Y*-axes), and in three-dimensional problems, three unit vectors are required for the *X*-, *Y*-, and *Z*-axes. The magnitude of each unit vector is equal to *one* (a *dimensionless* number). Usually the unit vector in the *X*-direction is labeled *i* and that in the *Y*-direction is labeled *j*. (Sometimes the symbols $\hat{i}$ and $\hat{j}$ are used.) The unit vectors, *i* and *j*, in an *X-Y* diagram are shown in Figure 5.10.

The advantage of using unit vectors depends on the fact that the product of a scalar and a vector is a vector (Equation 5.2). Therefore, multiplying a scalar by a unit vector gives *direction* to the product (the direction of the unit vector), but the magnitude is the same as that of the original scalar. That is, the product $F\,i$ has the direction of *i* (along the *X*-axis) and the magnitude of *F*.

Suppose that a vector **A** is described as follows: ''**A** lies along the *X*-axis and has a magnitude of 3 units.'' We can express this information regarding **A** in a simple way by using a unit vector. Figure 5.11(*a*) shows the situation, and the equation that reproduces the verbal statement above is

$$\mathbf{A} = 3\,i$$

FIGURE 5.10   The unit vectors *i* and *j*.

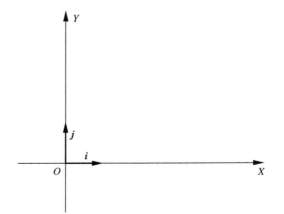

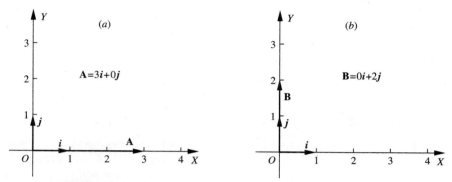

FIGURE 5.11   The representation of two vectors by using unit vectors.

Figure 5.11(*b*) shows a vector **B** two units in length which lies along the *Y*-axis. This vector is specified by the equation

$$\mathbf{B} = 2\,j$$

A vector that does not lie along one of the coordinate axes requires the use of both unit vectors for its specification. Figure 5.12 shows a vector **C** that makes an angle with the *X*-axis. We know that we can express this vector in terms of the component vectors, $\mathbf{C}_x$ and $\mathbf{C}_y$, as

$$\mathbf{C} = \mathbf{C}_x + \mathbf{C}_y \tag{5.17}$$

But each component vector lies along an axis, so we can write

$$\mathbf{C}_x = 4\,i, \quad \mathbf{C}_y = 3\,j \tag{5.18}$$

FIGURE 5.12   Resolving a vector **C** into components specified by unit vectors.

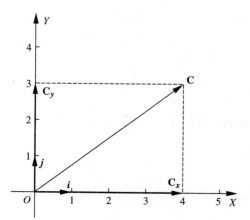

Combining Equations (5.17) and (5.18), the vector **C** is described by

$$\mathbf{C} = 4\,i + 3\,j \tag{5.19}$$

Any two-dimensional vector can be expressed in a similar way.

Because the addition of vectors involves adding components, as discussed in the preceding section, this operation is straightforward when using unit vectors. For example, if

$$\mathbf{A} = A_x i + A_y j$$
$$\mathbf{B} = B_x i + B_y j$$

the sum is

$$\mathbf{A} + \mathbf{B} = (A_x + B_x)\,i + (A_y + B_y)\,j \tag{5.20}$$

▶ **EXAMPLE 5.5.1** A force $\mathbf{F}_1$ exerts a 20-N pull on an object directed due north. A second force $\mathbf{F}_2$ pulls southeast on the same object with a force of 14.14 N. What is the net force on the object?

The force $\mathbf{F}_1$ is

$$\mathbf{F}_1 = 20\,j \tag{1}$$

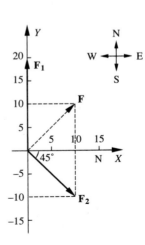

The force $\mathbf{F}_2$ does not lie along an axis and so it must first be resolved into components:

$$F_{2x} = F_2 \cos 45° = 14.14 \times 0.707 = 10 \text{ N}$$
$$F_{2y} = -F_2 \sin 45° = -14.14 \times 0.707 = -10 \text{ N}$$

where F_{2y} (which is directed *south*) is negative. Therefore,

$$\mathbf{F}_2 = 10\,i - 10\,j$$

The resultant force **F** is

$$\mathbf{F} = \mathbf{F}_1 + \mathbf{F}_2$$
$$= (0 + 10)\,i + (20 - 10)\,j$$
$$= 10\,i + 10\,j$$

This vector is shown by the dashed line in the diagram. Notice in the expression for **F** that the coefficients of *i* and *j* are both *positive*. This immediately tells us that

F lies in Quadrant I. Moreover, because the coefficients are equal in this case, **F** must make an angle of 45° with the *X*-axis—that is, the direction of **F** is northeast.

EXERCISES   *Use unit vectors to express the following statements:*

1. **F** represents a force of 33 N directed due north.
   (Ans. 12)
2. **x** represents a displacement of 10 m directed 45° south of due west.
   (Ans. 370)
3. **v** represents a velocity of 23 m/s directed due south.
   (Ans. 147)
4. **R** represents a resultant force of 10^4 N directed 30° south of due east.
   (Ans. 362)

5.6   SCALAR PRODUCT OF VECTORS

The multiplication of two scalar quantities, or even the multiplication of a scalar and a vector, is an easy process to understand. But how does one multiply two *vectors*? Because vectors have direction as well as magnitude, vector multiplication must be carefully defined. There are two important and different ways in which two vectors can be multiplied. The first of these is a process that produces a *scalar* result and is called the *scalar product*. The second type of multiplication produces a *vector* result and is called the *vector product* (see Section 5.7).

The scalar product can be understood in terms of the following physical situation. Suppose that a force **F** is applied to an object and as a result the object is displaced from a position *A* to a position *B;* this displacement is defined by the vector **s** (see Fig. 5.13). The *work* that is done by the force in causing the displacement is proportional to the component of **F** that is in the direction of the displacement **s**. The component of **F** that is *perpendicular* to **s** (in this case, the *vertical* component) causes no displacement because the motion is entirely horizontal; this component of **F** does no work. Therefore, the work done by the force is done entirely by the horizontal component of **F**:

Work done = (magnitude of displacement **s**)

$\times$ (component of force **F** in direction of **s**)   **(5.21)**

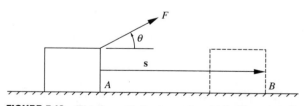

FIGURE 5.13   The force **F** displaces the object from *A* to *B*.

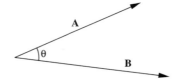

FIGURE 5.14   The scalar product of **A** and **B** is
$\mathbf{A} \cdot \mathbf{B} = AB \cos \theta$.

If θ is the angle between **s** and **F** (see Fig. 5.13), the component of **F** in the direction of **s** is

$$\text{component of } \mathbf{F} \text{ in direction of } \mathbf{s} = F \cos \theta \qquad (5.22)$$

Then, the work done by **F** is expressed as

$$W = Fs \cos \theta \qquad (5.23)$$

This type of multiplication of two vectors (**F** and **s**) is indicated in a shorthand notation by

$$W = \mathbf{F} \cdot \mathbf{s} \qquad (5.24)$$

The dot between **F** and **s** denotes the *scalar product* (the result W is a *scalar* quantity). Sometimes this product is called the *dot* product. The scalar product of two vectors always means the product of the *magnitudes* of the two vectors multiplied by the cosine of the angle between them. For two general vectors, **A** and **B**, as in Figure 5.14, the scalar product is

$$\mathbf{A} \cdot \mathbf{B} = |\mathbf{A}| \, |\mathbf{B}| \cos \theta = AB \cos \theta \qquad (5.25)$$

The scalar product is independent of the order of multiplication:

$$\mathbf{A} \cdot \mathbf{B} = \mathbf{B} \cdot \mathbf{A} \qquad (5.26)$$

That is, it does not matter whether we consider the product to be the component of **A** in the direction of **B** multiplied by $|\mathbf{B}|$ or the component of **B** in the direction of **A** multiplied by $|\mathbf{A}|$.

▶ **EXAMPLE 5.6.1**   If, in Figure 5.13 the force **F** is 50 N in a direction 30° above the horizontal (i.e., $\theta = 30°$) and if the horizontal displacement is 10 m, calculate the work done.

$$W = \mathbf{F} \cdot \mathbf{s} = Fs \cos \theta$$
$$= (50 \text{ N}) \times (10 \text{ m}) \times \cos 30°$$
$$= (500 \text{ N} \cdot \text{m}) \times 0.866$$
$$= 433 \text{ N} \cdot \text{m}$$

But 1 N·m is equal to 1 joule; therefore,

$$W = 433 \text{ J}$$

Let us now incorporate a unit vector description into the scalar product. In Figure 5.13 we can express the displacement **s** as

$$\mathbf{s} = x\,\boldsymbol{i} \tag{5.27}$$

where we choose the horizontal direction to be x and the vertical direction to be y, as usual. The components of **F** are

$$F_x = F \cos \theta$$
$$F_y = F \sin \theta$$

so that

$$\mathbf{F} = (F \cos \theta)\,\boldsymbol{i} + (F \sin \theta)\,\boldsymbol{j} \tag{5.28}$$

The scalar product of **F** and **s** is

$$W = \mathbf{F} \cdot \mathbf{s}$$
$$= [(F \cos \theta)\,\boldsymbol{i} + (F \sin \theta)\,\boldsymbol{j}] \cdot [x\,\boldsymbol{i}]$$
$$= Fx \cos \theta\,(\boldsymbol{i} \cdot \boldsymbol{i}) + Fx \sin \theta\,(\boldsymbol{j} \cdot \boldsymbol{i}) \tag{5.29}$$

What do the quantities $\boldsymbol{i} \cdot \boldsymbol{i}$ and $\boldsymbol{j} \cdot \boldsymbol{i}$ mean? Using Equation (5.25), we can write

$$\boldsymbol{i} \cdot \boldsymbol{i} = |\boldsymbol{i}|\,|\boldsymbol{i}| \cos \alpha$$

where the magnitude of the unit vector, $|\boldsymbol{i}|$, is *one;* α is the angle between the two unit vectors, but since the two vectors in this case are identical, α is zero and $\cos \alpha = 1$. Therefore

$$\boldsymbol{i} \cdot \boldsymbol{i} = 1 \tag{5.30}$$

Similarly, the angle β between $\boldsymbol{i}$ and $\boldsymbol{j}$ is 90° and $\cos 90° = 0$; therefore

$$\boldsymbol{j} \cdot \boldsymbol{i} = |\boldsymbol{j}|\,|\boldsymbol{i}| \cos \beta = 0 \tag{5.31}$$

Summarizing,

$$i \cdot i = 1$$
$$j \cdot j = 1$$
$$i \cdot j = 0$$
$$j \cdot i = 0 \tag{5.32}$$

Returning to Equation (5.29), we can now evaluate the expression for W:

$$W = Fx \cos \theta + 0 \tag{5.33}$$

which is just the result previously obtained (Equation 5.23).

EXERCISES

1. If $A = 2$ cm/s, $B = 3$ cm/s, and $\theta = 45°$, then $A \cdot B =$ _____ (Ans. 200)

2. If $A = 5$ N, $B = 25$ N, and $\theta = 90°$, then $A \cdot B =$ _____ (Ans. 123)

In each of the following cases take the direction of the displacement s to be along the positive X-axis and use the unit vector method.

3. If $F = 363$ N, $s = 21.6$ m, and $\theta = 90°$, then $F \cdot s =$ _____ (Ans. 292)

4. If $F = 363$ N, $s = 20$ m, and $\theta = 45°$, then $F \cdot s =$ _____ (Ans. 173)

5. If $F = 2 \times 10^6$ N, $s = 360$ m, and $\theta = 22°$, then $F \cdot s =$ _____ (Ans. 198)

5.7 THE VECTOR PRODUCT

The second type of vector multiplication gives a product which has both *magnitude* and *direction*, and hence is known as the *vector product*.

The vector product can be illustrated by means of the following physical example. Suppose that an object of mass m has instantaneous velocity **v**. Its *linear momentum* is then given by the expression $\mathbf{p} = m\mathbf{v}$.

Furthermore, suppose that the *displacement* of the object, relative to some origin O (see Fig. 5.15) is **r**. As illustrated in Figure 5.15 the two vectors **r** and **p** determine a plane, which we take to be the X-Y plane. In addition, θ denotes the angle between the vectors **r** and **p**. The Z-axis, of course, is perpendicular to the X-Y plane.

By definition, the *angular momentum* of the object relative to the origin O is equal to the *vector product* of **r** and **p**, which is written as **r** $\times$ **p** and is read as

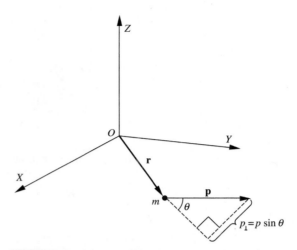

FIGURE 5.15   Vectors representing the displacement **r** and momentum **p** of a particle with mass m.

"**r** cross **p**." The term "cross" distinguishes the *vector* product from the *scalar* (or dot) product. It is customary to denote **r** × **p** by the vector **L**. That is,

$$\mathbf{r} \times \mathbf{p} = \mathbf{L}$$ (5.34)

The vector product **r** × **p** = **L** represents the angular momentum of the object, and has both magnitude and direction.

We first consider the *direction* of the angular momentum vector **L**. As illustrated in Figure 5.16, **L** = **r** × **p** is *perpendicular* to the plane formed by the vectors **r** and **p**. Furthermore, **L** = **r** × **p** points in the direction in which a *right-hand screw*

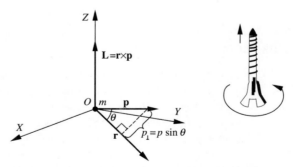

FIGURE 5.16   The vector product **r** × **p** = **L** represents the angular momentum of the mass relative to the origin O. The vector **L** = **r** × **p** points in the direction in which a right-hand screw would advance if the head of the screw is rotated from the first vector (**r**) toward the second vector (**p**) in the product. The magnitude of **L** is equal to $rp|\sin\theta|$.

would advance if the head of the screw is rotated from the first vector (**r**) toward the second vector (**p**). Therefore, the direction of **L** is along the positive Z-axis in Figure 5.16.

The *magnitude* of the angular momentum vector **L** = **r** × **p** is defined to be the magnitude of **r** times the magnitude of the component of **p** perpendicular to **r**. Referring to Figure 5.15 or Figure 5.16, we note that the component of **p** perpendicular to **r** is

$$p_\perp = p \sin \theta$$

where p is the magnitude of the vector **p**, and θ is the angle between **r** and **p**. Therefore, we find that the magnitude of **L** (which we denote by $|\mathbf{L}| = L$) is given by

$$L = r|p_\perp| = rp|\sin \theta| \tag{5.35}$$

where r denotes the magnitude of **r**, and $|\sin \theta|$ denotes the magnitude of $\sin \theta$. Evidently, $L = 0$ if **p** is *parallel* to **r**. Referring to Figure 5.15, we see that $\theta = 0°$ in this case. Therefore, for **p** parallel to **r**,

$$L = pr|\sin 0°| = 0$$

Similarly, if **p** is *antiparallel* to **r** ($\theta = 180°$), we also find

$$L = pr|\sin 180°| = 0$$

In contrast, if **p** is perpendicular to **r** ($\theta = 90°$), then

$$L = pr|\sin 90°| = pr$$

and L is a *maximum* for a given p and r if $\theta = 90°$.

▶ **EXAMPLE 5.7.1**   If, in Figure 5.15, $\theta = 30°$, $r = 10$ m, and $p = 30$ kg·m/s, what is the magnitude L of the angular momentum?

Since $\theta = 30°$, we find

$$\sin \theta = \sin 30° = 0.5$$

Substituting the values for r, p, and $\sin \theta$ into Equation (5.35) we find

$$L = (10 \text{ m}) \cdot (30 \text{ kg·m/s}) \cdot (0.5)$$
$$= 150 \text{ kg·m}^2/\text{s}$$

The vector product of two arbitrary vectors, **A** and **B**, is defined in a completely analogous manner. The magnitude and direction of the vector product

FIGURE 5.17   The vector product $\mathbf{A} \times \mathbf{B} = \mathbf{C}$ points in the direction which a right-hand screw would advance if the head of the screw is rotated from the first vector ($\mathbf{A}$) toward the second vector ($\mathbf{B}$) in the product. The magnitude of $\mathbf{C}$ is equal to $AB|\sin \theta|$.

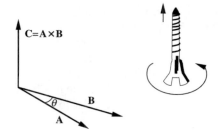

$$\mathbf{A} \times \mathbf{B} = \mathbf{C} \tag{5.36}$$

is illustrated in Figure 5.17.

The general prescription for determining the magnitude and direction of the vector product $\mathbf{A} \times \mathbf{B} = \mathbf{C}$ is summarized as follows:

> *Magnitude and direction of the vector product* $\mathbf{A} \times \mathbf{B} = \mathbf{C}$:
> The *direction* of $\mathbf{A} \times \mathbf{B} = \mathbf{C}$ is the direction which a right-hand screw would advance if the head of the screw is rotated from the first vector ($\mathbf{A}$) toward the second vector ($\mathbf{B}$) in the product (see Fig. 5.17). The *magnitude* of $\mathbf{A} \times \mathbf{B} = \mathbf{C}$ is given by
>
> $$C = AB \, |\sin \theta|$$
>
> where θ is the angle between $\mathbf{A}$ and $\mathbf{B}$, and A, B, and C denote the magnitudes of the vectors $\mathbf{A}$, $\mathbf{B}$, and $\mathbf{C}$, respectively.

It is important to note that the direction of a vector product *does* depend on the order in which the vectors are multiplied (this is in contrast to the scalar product of two vectors discussed in Section 5.6). Referring to Figure 5.17, it is clear that the direction of $\mathbf{B} \times \mathbf{A}$ is *opposite* to the direction of $\mathbf{A} \times \mathbf{B}$. This follows, since the direction which a right-hand screw would advance if the head of the screw is rotated from $\mathbf{B}$ toward $\mathbf{A}$ (rather than from $\mathbf{A}$ toward $\mathbf{B}$) is along $-\mathbf{C}$ in Figure 5.17. Since the *magnitude* of the vector product remains unchanged by the order of vector multiplication, we therefore conclude

$$\mathbf{B} \times \mathbf{A} = -\mathbf{A} \times \mathbf{B} \tag{5.37}$$

or equivalently,

$$\mathbf{A} \times \mathbf{B} = -\mathbf{B} \times \mathbf{A} \tag{5.38}$$

The unit vector notation can also be used to describe the vector product, but the details will not be discussed here.

EXERCISES

1. If **A** is directed due north, and **B** is directed due west, then the direction of **A** × **B** is _____
(Ans. 263)

2. If **A** is directed due east, and **B** is directed due south, then the direction of **A** × **B** is _____
(Ans. 272)

3. If **A** is directed 72° north of due east, and **B** is directed due west, then the direction of **B** × **A** is _____
(Ans. 345)

4. If **A** is directed due south, and **B** is directed due west, then the direction of **A** × **B** is _____
(Ans. 26)

5. If **A** is directed due south, and **B** is directed due west, then the direction of **B** × **A** is _____
(Ans. 154)

6. If $A = 10$ m, $B = 20$ m, and $\theta = 60°$, then $C = |$**A** × **B**$| = $ _____
(Ans. 234)

7. If $r = 10$ m, $p = 50$ kg·m/s, and $\theta = 5°$, then $L = $ _____
(Ans. 29)

8. If $r = 10$ m, $p = 50$ kg·m/s, and $\theta = 90°$, then $L = $ _____
(Ans. 252)

9. If $r = 10^6$ km, $p = 5 \times 10^{13}$ kg·m/s, and $\theta = 0°$, then $L = $ _____
(Ans. 223)

10. The distance of the Earth from the Sun is $r = 1.5 \times 10^8$ km, and the mean orbital speed of the Earth is $v = 30$ km/s. The mass of the Earth is 6.0×10^{24} kg, and its orbit about the Sun is approximately circular ($\theta = 90°$ in Fig. 5.15). What is the magnitude of the Earth's angular momentum around the Sun?
(Ans. 88)

5.8   TIME DERIVATIVES OF VECTORS

In problems involving motion, the vector that describes the position of a particle will change with time. Suppose that a particle moves uniformly along the positive X-axis in such a way that the distance from the origin to the particle increases by 25 m each second. At any time *t*, the position vector of the particle (that is, the vector which describes the displacement of the particle from the origin) is

$$\mathbf{s} = 25t\,\mathbf{i} \text{ meters} \tag{5.39}$$

where we take the time $t = 0$ to be the instant when the particle is at the origin.

The velocity of the particle is given by the derivative of the displacement with respect to time (Equation 4.15).* That is,

$$\mathbf{v} = \frac{d\mathbf{s}}{dt} \tag{5.40}$$

Notice that this is a *vector* equation—the velocity *vector* is equal to the time derivative of the displacement *vector*.

The unit vectors, *i* and *j*, are *constant* vectors—they do not change with time. Therefore, in Equation (5.39), *i* can be combined with the factor 25 into an overall constant coefficient of the time variable *t,* and we can write

$$\mathbf{s} = (25\,\mathbf{i})t \text{ meters} \tag{5.41}$$

---

*The displacement, velocity, and acceleration vectors are often written as **s**(*t*), **v**(*t*), and **a**(*t*), respectively, to indicate explicitly that these quantities depend, in the general case, upon the time *t*. For brevity, however, we adopt the short-form notation, **s**, **v** and **a**, throughout Sections 5.8 and 5.9.

The time derivative of the vector **s** is computed in exactly the same way as for scalars (see Equation 4.15). Therefore, since $dt/dt = 1$, we have

$$\mathbf{v} = \frac{d\mathbf{s}}{dt} = 25\,\boldsymbol{i}\text{ m/s} \qquad \text{(5.42)}$$

Thus, the velocity vector is a constant vector of 25 m/s in the positive X-direction.

The acceleration of a particle is given by the derivative of the velocity with respect to time (Equation 4.17). That is,

$$\mathbf{a} = \frac{d\mathbf{v}}{dt} \qquad \text{(5.43)}$$

which, again, is a vector equation.

If we consider a particle falling *downward* (call this the *positive Y*-direction), starting from rest at the origin, we have

$$\mathbf{s} = \frac{1}{2}gt^2\,\boldsymbol{j} \qquad \text{(5.44a)}$$

$$\mathbf{v} = \frac{d\mathbf{s}}{dt} = gt\,\boldsymbol{j} \qquad \text{(5.44b)}$$

$$\mathbf{a} = g\,\boldsymbol{j} \qquad \text{(5.44c)}$$

These equations are the vector equivalents of the familiar results.

In the following example, we treat a more complicated case of motion described by vectors.

▶ **EXAMPLE 5.8.1**   Suppose that a particle is given an initial horizontal velocity of 128 ft/s and then allowed to move freely under the influence of the Earth's gravity. How can we describe the position and the velocity of the particle as functions of time?

First, we note that the gravitational force acting on the particle is directed *downward* (we will call this the *positive Y*-direction). The initial velocity of the particle, however, is *horizontal* (we will call this the positive X-direction). Therefore, the motion separates into two independent parts (vector components). The horizontal motion takes place with constant speed because there is no force to change the velocity. The downward motion is uniformly accelerated due to gravity. Thus, we can write

$$x = v_{0x}t = 128t\text{ ft} \qquad \text{(1)}$$

$$y = \frac{1}{2}gt^2 = 16t^2\text{ ft} \qquad \text{(2)}$$

and the displacement vector is

$$\begin{aligned}
\mathbf{s} &= x\,\boldsymbol{i} + y\,\boldsymbol{j} \\
&= 128t\,\boldsymbol{i} + 16t^2\,\boldsymbol{j} \qquad \text{(3)}
\end{aligned}$$

The figure shows the parabolic path of this motion. Notice how the vector **s** changes with time.

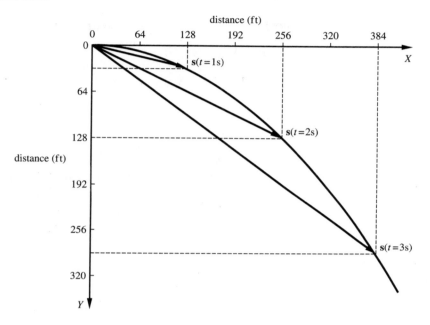

In order to find the velocity vector, we differentiate (3) with respect to t:

$$\mathbf{v} = \frac{d\mathbf{s}}{dt} = 128\,\boldsymbol{i} + 32t\,\boldsymbol{j} \tag{4}$$

Note that the velocity has a constant X-component and a Y-component that increases linearly with the time. The vector **v** is shown in the figure below for $t = 0$, 1, 2, and 3 s.

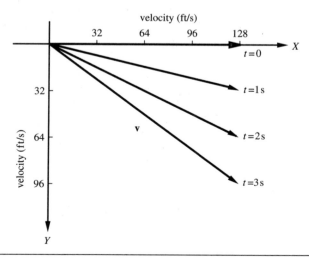

In the general case, a vector $\mathbf{A}(t)$ that varies with time t can be represented in component form as

$$\mathbf{A}(t) = A_x(t)\,\mathbf{i} + A_y(t)\,\mathbf{j} + A_z(t)\,\mathbf{k}$$

where $\mathbf{i}$, $\mathbf{j}$, and $\mathbf{k}$ are unit vectors along the X-, Y-, and Z-axes, respectively. The rule for differentiating $\mathbf{A}(t)$ with respect to t is:

$$\frac{d\mathbf{A}}{dt} = \frac{dA_x}{dt}\,\mathbf{i} + \frac{dA_y}{dt}\,\mathbf{j} + \frac{dA_z}{dt}\,\mathbf{k} \tag{5.45}$$

Note that the derivative of the vector $\mathbf{A}(t)$ with respect to t is also a vector.

In concluding this section we state, without proof, two important formulas from *vector* calculus. Consider two vectors, $\mathbf{A}(t)$ and $\mathbf{B}(t)$. The rule for differentiating the scalar product $\mathbf{A} \cdot \mathbf{B}$ is:

$$\frac{d}{dt}(\mathbf{A} \cdot \mathbf{B}) = \frac{d\mathbf{A}}{dt} \cdot \mathbf{B} + \mathbf{A} \cdot \frac{d\mathbf{B}}{dt} \tag{5.46}$$

The rule for differentiating the vector product $\mathbf{A} \times \mathbf{B}$ is:

$$\frac{d}{dt}(\mathbf{A} \times \mathbf{B}) = \frac{d\mathbf{A}}{dt} \times \mathbf{B} + \mathbf{A} \times \frac{d\mathbf{B}}{dt} \tag{5.47}$$

The student should verify Equations (5.46) and (5.47) for a few simple choices of $\mathbf{A}(t)$ and $\mathbf{B}(t)$.

▶ **EXAMPLE 5.8.2**   As illustrated in Figure 5.15, a particle with mass m moves in the X-Y plane with momentum $\mathbf{p}(t) = m\mathbf{v}(t)$ and displacement $\mathbf{r}(t)$ from the origin O. The corresponding angular momentum

$$\mathbf{L} = \mathbf{r} \times \mathbf{p} \tag{1}$$

is along the Z-axis as shown in Figure 5.16. Calculate an expression for the time rate of change of angular momentum $d\mathbf{L}/dt$.

From Equations (1) and (5.47), we obtain

$$\frac{d\mathbf{L}}{dt} = \frac{d\mathbf{r}}{dt} \times \mathbf{p} + \mathbf{r} \times \frac{d\mathbf{p}}{dt}$$

$$= \mathbf{v} \times m\mathbf{v} + \mathbf{r} \times m\frac{d\mathbf{v}}{dt} \tag{2}$$

Because **v** and *m***v** are *parallel* vectors, it follows that

$$\mathbf{v} \times m\mathbf{v} = 0 \tag{3}$$

on the right-hand side of Equation (2).

Similarly, from Newton's Second Law,

$$m \frac{d\mathbf{v}}{dt} = m\mathbf{a}$$

$$= \mathbf{F} \tag{4}$$

where **a** is the instantaneous acceleration, and **F** is the force exerted on mass m at point P in Figure 5.15. Equation (2) then reduces to

$$\frac{d\mathbf{L}}{dt} = \mathbf{r} \times \mathbf{F} \tag{5}$$

The vector

$$\boldsymbol{\tau} = \mathbf{r} \times \mathbf{F} \tag{6}$$

occurring in Equation (5) is referred to as the *torque* exerted on the mass m by the force **F** about the origin O.

We denote the magnitudes of the vectors $\boldsymbol{\tau}$, **r**, and **F** by τ, r, and F, respectively, and we denote the angle between the vectors **r** and **F** by α. Then the *magnitude* of the torque is

$$\tau = rF \left| \sin \alpha \right| \tag{7}$$

Note that if the displacement **r** and the force **F** are parallel or antiparallel, then the torque τ is equal to zero because $\sin 0° = 0$ or $\sin 180° = 0$. In general, because the vectors **r** and **F** are in the X-Y plane for the example considered in Figure 5.15, we note from Equation (6) that the torque vector $\boldsymbol{\tau}$ is along the positive or negative Z-axis.

Finally the torque τ should be thought of as a measure of the accelerating "spin-up" or decelerating "spin-down" of the angular motion of the mass m about the origin O in Figure 5.15.

▶ **EXAMPLE 5.8.3**   As illustrated in Figure 5.8.3 accompanying this example, a uniform-density ladder of length ℓ and weight $W = mg$ leans against a smooth (frictionless) wall, making an angle θ with the horizontal. A frictional force $\mathbf{F}_f$, parallel to the ground, acts on the base of the ladder as shown in the figure and prevents the ladder from slipping.

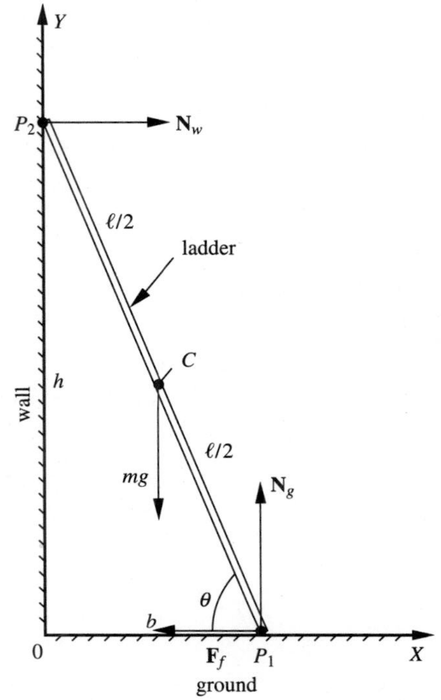

FIGURE 5.8.3 A ladder of length ℓ and weight mg leans against a smooth wall and makes an angle θ with the horizontal.

(a) Determine an expression for the frictional force $\mathbf{F}_f$ in terms of ℓ, θ and the weight mg of the ladder.

(b) If the coefficient of static friction between the ladder and the ground is $\mu_s = 0.5$, calculate the minimum angle θ_{min} at which the ladder can be placed such that the ladder will not slip.

In answer to part (a), the forces acting on the ladder are illustrated in the figure. Because the wall is frictionless, there is no force parallel to the surface of the wall; however, there is a force $\mathbf{N}_w$ *normal* to the wall at the point of contact. At the base of the ladder, the frictional force $\mathbf{F}_f$ is inward (toward the wall), and the force $\mathbf{N}_g$ is *normal* to the ground at the point of contact. Because the ladder is uniform, the center of mass is a distance $\ell/2$ along its length, and the gravitational force associated with the weight mg of the ladder acts downward at the center-point C shown in the figure. By geometric construction the distances h and b in the figure are related to θ and ℓ by

$$h = \ell \sin \theta$$
$$b = \ell \cos \theta \tag{1}$$

where ℓ and θ are specified quantities.

Because the ladder is stationary, the net force on the ladder in the vertical direction and in the horizontal direction must be exactly zero. Therefore, from the figure

$$N_g + (-mg) = 0$$
$$N_w + (-F_f) = 0$$

or equivalently,

$$N_g = mg \tag{2}$$

and

$$F_f = N_w \tag{3}$$

While Equations (2) and (3) are true statements, they are not adequate to determine a closed expression for the frictional force F_f in terms of ℓ, θ, and mg. To do this, we make use of the fact that the net *torque* τ on the ladder must be exactly zero; otherwise, the ladder would be rotating about some point such as O. Of course, this would be contradictory to the assumption that the ladder is in *static* equilibrium.

We calculate the total torque τ on the ladder about the point O by making use of the definition of torque in Equation (6) of Example 5.8.2, and summing over all of the forces acting on the ladder according to

$$\tau = \sum_i \mathbf{r}_i \times \mathbf{F}_i \tag{4}$$

From Figure 5.8.3, the positive Z-axis is *out of the page* and $\mathbf{r}_i$ and $\mathbf{F}_i$ are in the X-Y plane. Therefore, for each of the force components, the individual torque contributions $\mathbf{r}_i \times \mathbf{F}_i$ are along the positive or negative Z-axis. Introducing the unit vector i along the X-axis and the unit vector j along the Y-axis (see Section 5.5), and referring to the figure, we obtain for the torque τ exerted on the ladder about the point O

$$\tau = h\mathbf{j} \times N_w \mathbf{i} + [(b/2)\mathbf{i} + (h/2)\mathbf{j}] \times (-mg)\mathbf{j} + b\mathbf{i} \times [N_g\mathbf{j} + (-F_f)\mathbf{i}] \tag{5}$$

Here, we have represented the displacement vector from O to P_2 by $h\mathbf{j}$, from O to C by $(b/2)\mathbf{i} + (h/2)\mathbf{j}$, and from O to P_1 by $b\mathbf{i}$. The student can readily verify from the definition of cross-product that

$$i \times i = 0$$
$$j \times j = 0 \tag{6}$$

and

$$j \times i = -k$$
$$i \times j = k \tag{7}$$

where **k** is a unit vector along the *positive* Z-axis. (Refer to Figure 5.16 and the discussion of unit vectors in Section 5.5.) Making use of Equations (6) and (7), the expression for the net torque τ in Equation (5) readily reduces to

$$\tau = \left[-hN_w - \frac{b}{2}mg + bN_g \right]\mathbf{k} \tag{8}$$

However, because the ladder is *stationary* and *not* rotating, the net torque in Equation (8) is equal to zero

$$\tau = 0 \tag{9}$$

which gives

$$-hN_w - \frac{b}{2}mg + bN_g = 0 \tag{10}$$

We now make use of Equations (2) and (3) to eliminate $N_g = mg$ and $N_w = F_f$ in Equation (10). This gives

$$-hF_f - \frac{b}{2}mg + bmg = 0 \tag{11}$$

Solving for F_f, we obtain

$$F_f = \frac{1}{2}mg\,\frac{b}{h} \tag{12}$$

or equivalently

$$F_f = \frac{1}{2}mg \cot \theta \tag{13}$$

where use is made of Equation (1). Equation (13) is the desired expression for the frictional force F_f in answer to part (a). A striking feature of Equation (13) is that F_f depends on mg and θ but *not* on the length ℓ of the ladder.

In answer to part (b), recall that the frictional force F_f satisfies

$$F_f \leq \mu_s N_g \tag{14}$$

where μ_s is the coefficient of static friction, and $N_g = mg$ is the force normal to the ground at the point of contact (see the figure). When the ladder is on the verge of slipping, the force of friction must be a maximum with

$$(F_f)_{max} = \mu_s N_g \tag{15}$$

Therefore, to determine $\theta_{\min}$, we set $F_f = \mu_s N_g$ in Equation (13), which gives

$$\mu_s N_g = \frac{1}{2} mg \cot \theta_{\min} \tag{16}$$

But $N_g = mg$ from Equation (2). Therefore, Equation (16) gives

$$\mu_s mg = \frac{1}{2} mg \cot \theta_{\min}$$

or equivalently

$$\cot \theta_{\min} = 2 \mu_s \tag{17}$$

That is, the angle $\theta_{\min}$ at which slipping begins to occur depends only on μ_s, the coefficient of static friction. For the special case where $\mu_s = 0.5$, Equation (17) gives $\cot \theta_{\min} = 1$, or

$$\theta_{\min} = 45°$$

Therefore, for this example, the base will slide unless the ladder is inclined at angles greater than 45°.

5.9   UNIFORM CIRCULAR MOTION

We can now bring together several ideas discussed in previous sections to give a vector description of uniform circular motion. From Section 2.11 we know that the rectangular coordinates (x, y) of the position of a particle described by the polar coordinates (r, θ) are (see Equation 2.32)

$$x = r \cos \theta$$
$$y = r \sin \theta \tag{5.48}$$

These coordinates are illustrated in Figure 5.18. Furthermore, from the discussion in Section 2.6, we know that a particle undergoing uniform circular motion has a constant angular speed ω. If the particle passes through $\theta = 0$ at $t = 0$, we can write the angular coordinate for any later time t as

$$\theta = \omega t \tag{5.49}$$

Combining Equations (5.48) and (5.49), we have

$$x = r \cos \omega t$$
$$y = r \sin \omega t \tag{5.50}$$

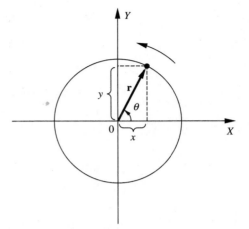

FIGURE 5.18　Coordinates for uniform circular motion.

where r is a constant for motion in a circle. Note from Equation (5.50) that the particle's motion projected onto the X- or Y-axis is *cyclic* (or *periodic*). Since the x- and y-coordinates of the particle are proportional to cos ωt and sin ωt, the *simple harmonic motion* described by Equation (5.50) has the period

$$T = \frac{2\pi}{\omega}$$

We can now write down the vector that describes the position of the particle as a function of time:*

$$\mathbf{r} = x\boldsymbol{i} + y\boldsymbol{j}$$
$$= (r \cos \theta)\boldsymbol{i} + (r \sin \theta)\boldsymbol{j}$$
$$= (r \cos \omega t)\boldsymbol{i} + (r \sin \omega t)\boldsymbol{j} \tag{5.51}$$

By referring to Figure 5.18, we can see that $\mathbf{r}$ is a vector that rotates uniformly around the origin.

The velocity of the particle can be obtained by differentiating $\mathbf{r}$ with respect to t:

$$\mathbf{v} = \frac{d\mathbf{r}}{dt}$$
$$= (-r\omega \sin \omega t)\boldsymbol{i} + (r\omega \cos \omega t)\boldsymbol{j} \tag{5.52}$$

*For problems that involve polar coordinates it is customary to use the vector $\mathbf{r}$ (rather than s) to denote the position of the particle.

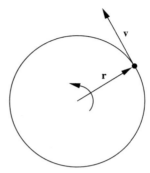

FIGURE 5.19   For circular motion, **r** is always perpendicular to **v**.

How do we interpret this result? We could construct a set of vectors **v** for various times and compare them with the corresponding vectors **r**. But it is more interesting to examine the scalar product of **r** and **v**. Using Equations (5.51) and (5.52), we have

$$\mathbf{r} \cdot \mathbf{v} = [(r \cos \omega t)\,i + (r \sin \omega t)\,j] \cdot [(-r\omega \sin \omega t)\,i + (r\omega \cos \omega t)\,j]$$

$$= (-r^2\omega \cos \omega t \sin \omega t)\,i \cdot i$$

$$+ (r^2\omega \cos^2 \omega t)\,i \cdot j$$

$$+ (-r^2\omega \sin^2 \omega t)\,j \cdot i$$

$$+ (r^2\omega \sin \omega t \cos \omega t)\,j \cdot j \qquad \text{(5.53)}$$

In this equation, the middle two terms on the right-hand side are each zero because $i \cdot j = j \cdot i = 0$ (Equations 5.32). Furthermore, $i \cdot i = j \cdot j = 1$, so the first and last terms exactly cancel. We therefore find

$$\mathbf{r} \cdot \mathbf{v} = 0 \qquad \text{(5.54)}$$

This result means that the scalar product is *always* zero (independent of the time); this is possible only if the vectors **r** and **v** are always *perpendicular*. Because the motion takes place in a circle, we know this to be true (see Figure 5.19).

Next, let us compute the acceleration of the particle by taking the time derivative of **v** (Equation 5.52). We find

$$\mathbf{a} = \frac{d\mathbf{v}}{dt}$$

$$= (-r\omega^2 \cos \omega t)\,i + (-r\omega^2 \sin \omega t)\,j \qquad \text{(5.55)}$$

Comparing this expression for **a** with Equation (5.51) for **r**, we see that we can write

$$\mathbf{a} = -\omega^2\,\mathbf{r} \qquad \text{(5.56)}$$

The negative sign in this equation means that the acceleration vector **a** always has a direction opposite to that of the position vector **r**. Thus, **a** is always directed toward the *center* of the circular path. For this reason, **a** is called the *centripetal* (''center-seeking'') acceleration.

From Equation (5.51) we can calculate the *magnitude* of the displacement vector **r**(t) from an evaluation of

$$\mathbf{r} \cdot \mathbf{r} = [(r \cos \omega t)\,\boldsymbol{i} + (r \sin \omega t)\,\boldsymbol{j}] \cdot [(r \cos \omega t)\,\boldsymbol{i} + (r \sin \omega t)\,\boldsymbol{j}] \tag{5.57}$$

Because $\boldsymbol{i} \cdot \boldsymbol{j} = \boldsymbol{j} \cdot \boldsymbol{i} = 0$ and $\boldsymbol{i} \cdot \boldsymbol{i} = \boldsymbol{j} \cdot \boldsymbol{j} = 1$, this readily gives

$$\begin{aligned}
\mathbf{r} \cdot \mathbf{r} &= \mathbf{r}^2 \cos^2 \omega t + \mathbf{r}^2 \sin^2 \omega t \\
&= \mathbf{r}^2 (\cos^2 \omega t + \sin^2 \omega t)
\end{aligned} \tag{5.58}$$

But $\cos^2 \omega t + \sin^2 \omega t = 1$. Therefore, Equation (5.57) reduces to the expected result

$$\mathbf{r} \cdot \mathbf{r} = r^2 \tag{5.59}$$

That is the length of the displacement vector is $|\mathbf{r}| = r$, the constant radius of the circle in Figure 5.18.

In a similar manner (see Exercise 1 at the end of Section 5.9) it can be shown that

$$\mathbf{v} \cdot \mathbf{v} = \omega^2 \, r^2 \tag{5.60}$$

Therefore, the *speed* of the particle is $v = \sqrt{\mathbf{v} \cdot \mathbf{v}}$, or from Equation (5.60)

$$v = \omega r \tag{5.61}$$

Another interesting property of the motion is the angular momentum

$$\mathbf{L} = \mathbf{r} \times \mathbf{p} \tag{5.62}$$

of the particle about the origin in Figure 5.18. Here, $\mathbf{p} = m\mathbf{v}$ is the momentum, and the angular momentum **L** is defined in Equation (5.34). To determine **L** it is possible to make use of the definitions of **r** and $\mathbf{p} = m\mathbf{v}$ in Equations (5.51) and (5.52) and evaluate all of the component contributions of $\mathbf{r} \times \mathbf{p}$ term by term. A simpler approach is to make direct use of the fact that **r** and **v** are both in the *X-Y* plane and are *perpendicular* to one another as shown in Equation (5.54) and illustrated in Figure 5.19. Therefore, comparing Figure 5.16 and Figure 5.19, the *direction* of $\mathbf{r} \times m\mathbf{v}$ is out of the page in Figure 5.18, that is along the *positive Z*-axis. Moreover, the magnitude of the angular momentum vector **L** is

$$L = r\,mv\,|\sin 90°| \tag{5.63}$$

because **r** and **v** are perpendicular. Because $v = \omega r$ and $\sin 90° = 1$, Equation (5.63) gives

$$L = m\omega r^2$$

for the magnitude of the angular momentum about the origin in Figure 5.18.

▶ **EXAMPLE 5.9.1** For the case of uniform circular motion considered in Section 5.9, calculate the torque

$$\boldsymbol{\tau} = \mathbf{r} \times \mathbf{F} \qquad (1)$$

exerted on the particle in its motion about the origin in Figure 5.18.

From Newton's Second Law, $\mathbf{F} = m\mathbf{a}$ where m is the mass of the particle and $\mathbf{a}$ is the instantaneous acceleration determined in Equation (5.55). Therefore, Equation (1) can be expressed as

$$\boldsymbol{\tau} = m\mathbf{r} \times \mathbf{a} \qquad (2)$$

The vector product in Equation (2) can be evaluated by direct calculation. However, from Equations (5.51) and (5.56), it was shown that

$$\mathbf{a} = -\omega^2 \mathbf{r} \qquad (3)$$

Therefore, Equation (2) can be expressed as

$$\boldsymbol{\tau} = m\mathbf{r} \times (-\omega^2 \mathbf{r})$$
$$= -m\omega^2 \mathbf{r} \times \mathbf{r} \qquad (4)$$

In Section 5.7 it was shown that the vector product of parallel (or antiparallel) vectors is equal to zero. Therefore $\mathbf{r} \times \mathbf{r} = 0$ in Equation (4), which gives

$$\boldsymbol{\tau} = 0 \qquad (5)$$

Therefore, for the case of uniform circular motion where the speed of the particle is *constant,* the torque experienced by the particle in Figure 5.18 is *zero.* This is not surprising because a non-zero torque would cause the particle to speed up or slow down as it moves around the circle in Figure 5.18. (See Example 5.8.2.)

EXERCISES

1. Find the magnitude of the velocity of the particle (that is, the *speed*) by taking the scalar product of $\mathbf{v}$ (Equation 5.52) with itself, and then taking the square root: $v = \sqrt{\mathbf{v} \cdot \mathbf{v}}$. (Ans. 18)

2. Compute the scalar product of $\mathbf{r}(t = 0)$ and $\mathbf{r}(t = \frac{3}{2} T)$ where $T = 2\pi/\omega$ is the period of the motion. Interpret the result. (Ans. 339)

A particle of mass m is constrained to move on a circle with constant radius r as shown in Figure 5.18. However, in this example, the particle does not move with constant speed; rather the azimuthal motion accelerates at a constant rate according to

$$\theta = \alpha t^2$$

3. Evaluate the instantaneous coordinates $x(t)$ and $y(t)$.

(Ans. 385)

4. Evaluate the instantaneous velocity $\mathbf{v}(t) = d\mathbf{x}/dt$.

(Ans. 138)

5. Evaluate the scalar product $\mathbf{r}(t) \cdot \mathbf{v}(t)$.   (Ans. 113)

6. Evaluate the angular momentum $\mathbf{L} = \mathbf{r} \times m\mathbf{v}$.

(Ans. 204)

7. Evaluate the instantaneous acceleration $\mathbf{a}(t) = d\mathbf{v}/dt$. Express your answer in terms of t, $\mathbf{r}(t)$, and $\mathbf{v}(t)$.

(Ans. 67)

8. Evaluate the torque $\tau = \mathbf{r} \times m\mathbf{a}$.   (Ans. 306)

CHAPTER

6

Integral Calculus

6.1 INDEFINITE INTEGRALS

The operation that is inverse to differentiation is called *integration*. If we differentiate a function of t, $I(t)$, we obtain another function of t, which we denote by $f(t)$:

$$f(t) = \frac{dI}{dt} \tag{6.1}$$

In differential calculus, we are given the function $I(t)$ and asked to find $f(t)$. But suppose, instead, that $f(t)$ is given and we are asked to determine $I(t)$. The procedure can be illustrated with a few simple examples.

▶ **EXAMPLE 6.1.1** Suppose that $f(t) = b$, where b is a constant. What function $I(t)$ is related to this $f(t)$ by Equation (6.1)?

Referring to Table 4.1, we see that the derivative of $I(t) = bt$ is equal to the constant b. But this is not the most general solution to the stated problem. The reason is that the derivative of *any* constant C is equal to *zero*. Therefore, the derivative of

$$I(t) = bt + C \tag{1}$$

is also equal to b. That is,

$$\frac{dI}{dt} = \frac{d}{dt}(bt + C) = b + 0 \tag{2}$$

and $I(t)$ in (1) is the most general solution to the problem.

▶ **EXAMPLE 6.1.2**   Suppose that $f(t) = bt^2$, where b is a constant. What function $I(t)$ is related to this $f(t)$ by Equation (6.1)?

Comparing

$$\frac{dI}{dt} = bt^2 \qquad (1)$$

with the entries in Table 4.1 suggests that $I(t)$ is proportional to t^3. Trial and error shows that the proportionality constant is $b/3$. As before, we must add an arbitrary constant C in order to obtain the most general solution:

$$I(t) = \frac{1}{3} bt^3 + C \qquad (2)$$

Checking,

$$\frac{dI}{dt} = \frac{d}{dt}\left(\frac{1}{3} bt^3 + C\right) = bt^2 \qquad (3)$$

In the examples above, the function $I(t)$ is called the *indefinite integral* of $f(t)$. The term "indefinite" is used because $I(t)$ is defined only to within an additive constant C.

The function $I(t)$ that satisfies Equation (6.1) for a given $f(t)$ is written as

$$I(t) = \int f(t)\, dt \qquad (6.2)$$

and means that $I(t)$ is the *indefinite integral* of $f(t)$. The function $f(t)$ is called the *integrand*, and the symbols $\int \cdots dt$ stand for "calculate the indefinite integral of the integrand with respect to t." Using this notation for the results of Examples 6.1.1 and 6.1.2, we can write

$$\int b\, dt = bt + C \qquad (6.3a)$$

$$\int bt^2\, dt = \frac{1}{3} bt^3 + C \qquad (6.3b)$$

The constant C is called the *constant of integration*. In general, for $f(t) = bt^n$ and $n \neq -1$, it can be shown that (see Example 6.1.3)

$$\int bt^n\, dt = \frac{1}{n+1} bt^n + C, \text{ for } n \neq -1 \qquad (6.3c)$$

▶ **EXAMPLE 6.1.3** Consider the case where $f(t) = bt^n$, where $n \neq -1$ and b is a constant. What function $I(t)$ is related to this $f(t)$ by $dI/dt = f$? Comparing

$$\frac{dI}{dt} = bt^n \tag{1}$$

with the entries in Table 4.1 indicates that $I(t)$ is proportional to t^{n+1}. Trial and error shows that the proportionality constant is $b/(n + 1)$ provided $n \neq -1$. Therefore the most general solution is

$$I(t) = \frac{b}{n + 1} t^{n+1} + C \tag{2}$$

where C is an arbitrary constant. Checking, we obtain

$$\frac{dI}{dt} = \frac{d}{dt}\left(\frac{b}{n + 1} t^{n+1} + C\right)$$

$$= \frac{b}{n + 1}(n + 1)t^n$$

$$= bt^n$$

as required.

Some of the indefinite integrals often encountered in physics problems are listed in Table 6.1. It is easy to verify that every pair of entries, $f(t)$ and $I(t)$, satisfy Equation (6.1).

It is important to note that integration, just as differentiation, is a *linear* operation. That is,

$$\int [\alpha f(t) + \beta g(t)]\, dt = \alpha \int f(t)\, dt + \beta \int g(t)\, dt \tag{6.4}$$

where α and β are constants.

In physical applications the constant of integration C is determined by the conditions that hold at some particular time (or place)—these are called the *initial* conditions (or *boundary* conditions). The evaluation of integration constants can be illustrated by an example familiar from the study of the motion of a particle in one dimension (Section 3.4).

TABLE 6.1 Some Indefinite Integrals

$f(t)$	$I(t) = \int f(t)\, dt$
0	C
$\alpha = $ constant	$\alpha t + C$
t	$\left(\frac{1}{2}\right)t^2 + C$
t^2	$\left(\frac{1}{3}\right)t^3 + C$
t^3	$\left(\frac{1}{4}\right)t^4 + C$
$\sqrt{t}$	$\left(\frac{2}{3}\right)t^{3/2} + C$
$1/t$	$\ln t + C$
$1/t^2$	$-1/t + C$
$1/t^3$	$-1/(2t^2) + C$
$1/\sqrt{t}$	$2\sqrt{t} + C$
$t^n \ (n \neq -1)$	$(n+1)^{-1} t^{n+1} + C$
$\sin t$	$-\cos t + C$
$\sin \alpha t$	$-\alpha^{-1} \cos \alpha t + C$
$\cos t$	$\sin t + C$
$\cos \alpha t$	$\alpha^{-1} \sin \alpha t + C$
$\tan t$	$-\ln |\cos t| + C$
$\tan \alpha t$	$-\alpha^{-1} \ln |\cos \alpha t| + C$
e^t	$e^t + C$
$e^{\alpha t}$	$\alpha^{-1} e^{\alpha t} + C$
e^{-t}	$-e^{-t} + C$
$e^{-\alpha t}$	$-\alpha^{-1} e^{-\alpha t} + C$
$\ln t$	$t \ln t - t + C$
$\ln \alpha t$	$t \ln \alpha t - t + C$

▶ **EXAMPLE 6.1.4** A particle of mass m moves along the X-axis and experiences a constant force F_0 in the positive x-direction. At time $t = 0$, the displacement and velocity are $x(t = 0) = x_0$ and $v(t = 0) = v_0$, where x_0 and v_0 are constants. (These are the *initial conditions*.) Determine the equations that describe the motion.

From Newton's Second Law, the acceleration a of the particle is

$$a = \frac{F_0}{m} = \text{constant} \tag{1}$$

The velocity $v(t)$ is related to the acceleration by

$$\frac{dv}{dt} = a \tag{2}$$

This equation is a *differential equation* for the velocity $v(t)$. Because a is a constant, the second entry in Table 6.1 shows that the solution of (2) is

$$v(t) = at + C \tag{3}$$

To determine the integration constant C we make use of the initial condition, $v(t = 0) = v_0$. Evaluating (3) at $t = 0$, we find

$$v(t = 0) = v_0 = 0 + C \tag{4}$$

That is, $C = v_0$, and (3) becomes

$$v(t) = v_0 + at \tag{5}$$

which is the familiar result.

The displacement x is related to $v(t)$ by

$$\frac{dx}{dt} = v(t) \tag{6}$$

Combining (5) and (6), we have the *differential equation* for $x(t)$:

$$\frac{dx}{dt} = v_0 + at \tag{7}$$

Using the second and third entries in Table 6.1, (7) can be integrated to give

$$x(t) = v_0 t + \frac{1}{2} at^2 + C' \tag{8}$$

The initial condition $x(t = 0) = x_0$ is used to evaluate C':

$$x(t = 0) = x_0 = 0 + 0 + C' \tag{9}$$

Thus, $C' = x_0$, and (8) reduces to the well-known equation for uniformly accelerated motion,

$$x(t) = x_0 + v_0 t + \frac{1}{2} at^2 \tag{10}$$

The example just completed is instructive for two important reasons. First, we have *derived from basic principles* the correct expressions for $x(t)$ and $v(t)$ for the case of uniformly accelerated motion. Second, we have illustrated the method for determining integration constants from the initial conditions.

In the example above we considered the case of constant acceleration. But, in more complex situations, the force and the acceleration can vary with time. Then, we have a differential equation of the form,

$$\frac{dv}{dt} = a(t) \tag{6.5}$$

which can be integrated to give

$$v(t) = \int a(t)\, dt \qquad\qquad \textbf{(6.6)}$$

For many functions $a(t)$, the integral in Equation (6.6) can be evaluated and the initial conditions used to determine the constant of integration.* Then, the displacement can be obtained from

$$x(t) = \int v(t)\, dt \qquad\qquad \textbf{(6.7)}$$

with the initial conditions used to evaluate the integration constant. Some examples of time-dependent forces $F(t)$ for which $v(t)$ and $x(t)$ can be found by simple integration are included in the following exercises.

EXERCISES   *Determine I(t) for the following f(t):*

1. $f(t) = t^{5/2}$ (Ans. 129)
2. $f(t) = 6t^{-5/2}$ (Ans. 373)
3. $f(t) = 3t^2 + 8t^9$ (Ans. 15)
4. $f(t) = 6 \sin 6t$ (Ans. 187)

5. $f(t) = t^2 - \ln 6t$ (Ans. 265)
6. $f(t) = \tan 3t - 1/t$ (Ans. 63)
7. $f(t) = te^t$ (Ans. 363)
8. $f(t) = e^{-\alpha t} - e^{\alpha t}$ (Ans. 85)

A particle of mass m moves along the X-axis subject to a time-dependent force F(t). The initial conditions are $x(t = 0) = x_0$ and $v(t = 0) = v_0$. Determine a(t), v(t), and x(t) for the following F(t):

9. $F(t) = F_0 e^{2t}$, where F_0 = constant. (Ans. 105)
10. $F(t) = \alpha t^2$, where α = constant. (Ans. 380)

11. $F(t) = F_0 \sin t$, where F_0 = constant. (Ans. 218)
12. $F(t) = \alpha t^{1/3}$, where α = constant. (Ans. 3)

6.2   DEFINITE INTEGRALS

As discussed in Section 4.2, *differentiation* has a simple geometrical interpretation—the derivative of a function $f(t)$ is equal to the *slope* of the $f(t)$ *versus* t curve. *Integration* also has a simple geometrical interpretation—the integral of a function $f(t)$ is related to the *area* under the $f(t)$ *versus* t curve. We will begin by *defining* the area under the $f(t)$ curve to be the *definite integral* of $f(t)$. We will then examine some of the properties of the integral defined in this way, and in the following section we will demonstrate the simple relationship between *indefinite* and *definite* integrals.

Consider the four different functions $f(t)$ illustrated in Figures 6.1*a, b, c,* and *d.*

---

*In the event that $a(t)$ is a complicated function, the integral may require numerical methods for evaluation. Not all functions can be integrated in simple ways!

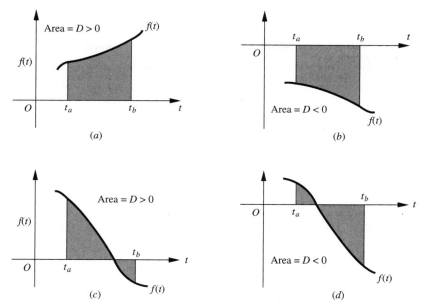

FIGURE 6.1 The areas under curves defined by various functions $f(t)$. The areas above the t-axis are defined to be *positive,* and those below the t-axis are defined to be *negative.* A particular area can include both positive and negative portions.

The *area under the curve* (the shaded area) in each case means the area bounded by the t-axis, the curve $f(t)$, and the two straight lines, $t = t_a$ and $t = t_b$. The area under the curve of a positive function (Fig. 6.1*a*) is defined to be *positive,* whereas the area under (actually, *above*) a negative function (Fig. 6.1*b*) is defined to be negative. If the function takes on both positive and negative values in the region of interest (Figs. 6.1*c* and *d*), the positive and negative areas add algebraically. In Figure 6.1*c,* for example, the net result is a *positive* area, whereas in Figure 6.1*d,* the net result is a *negative* area.

The total area under the curve of $f(t)$, between the *definite* limits t_a and t_b, as shown in Figure 6.1, is called the *definite integral* of $f(t)$ from $t = t_a$ to $t = t_b$. The notation used for the definite integral is

$$\text{Area} = D = \int_{t_a}^{t_b} f(t)\, dt \tag{6.8}$$

Notice that D is a *definite number* and is *not* a function of t as is the indefinite integral $I(t)$, exhibited in Equation (6.2). We see that there is a similarity in the notation for D and $I(t)$, and this is suggestive of a relationship between the definite and indefinite integrals. This connection will be demonstrated in Section 6.3.

Although we have identified the definite integral as an *area,* the dimensions (or units) of D need not correspond to the conventional dimensions of m², cm², ft², and so forth. From Figure 6.1 or Equation (6.8), we see that the units of D are

$$[\text{Units of } D] = [\text{Units of } f(t)] \times [\text{Units of } t] \tag{6.9}$$

For example, if $f(t)$ represents a velocity and t represents the time, the units of D are (m/s) $\times$ (s) $=$ m. Thus, the "area" under the v-t curve has the units of *displacement*. On the other hand, if $f(t)$ represents an acceleration and t represents the time, the units of D are (m/s²) $\times$ (s) $=$ m/s, the units of *velocity*.

Some definite integrals can be evaluated directly from the area definition of D. We now consider two such examples.

▶ **EXAMPLE 6.2.1**   The function $f(t) = \alpha = $ constant is illustrated in the graph below. What is the definite integral of $f(t)$ between the limits $t = t_a$ and $t = t_b$?

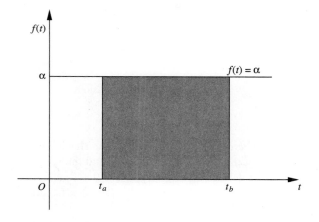

According to Equation (6.8), the definite integral is

$$D = \int_{t_a}^{t_b} f(t)\, dt = \int_{t_a}^{t_b} \alpha\, dt \tag{1}$$

For the shaded rectangle in the diagram, the area under the curve is

$$\text{Area} = D = (\text{base}) \times (\text{height}) = (t_b - t_a) \times \alpha \tag{2}$$

so that the definite integral of $f(t) = \alpha$ from $t = t_a$ to $t = t_b$ is

$$D = \alpha(t_b - t_a) \tag{3}$$

▶ **EXAMPLE 6.2.2** A particle, accelerated uniformly from rest, has a velocity $v(t) = at$, where the acceleration is $a = 10$ m/s^2. What is the definite integral of $v(t)$ from $t = 0$ to $t = 10$ s?

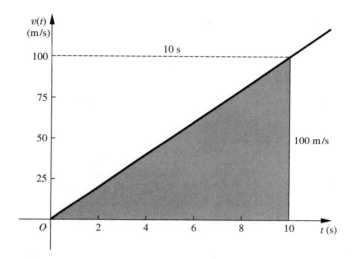

For this example, the integration limits are $t_a = 0$ and $t_b = 10$ s. Therefore, the definite integral is

$$D = \int_0^{10\text{s}} v(t)\,dt = \int_0^{10\text{s}} a\,t\,dt \qquad (1)$$

This integral is equal to the shaded area under the straight line representing $v(t) = at$ in the diagram above. That is,

$$\text{Area} = D = \frac{1}{2}(\text{base}) \times (\text{height})$$

$$= \frac{1}{2}(10\text{ s}) \times (100\text{ m/s})$$

$$= 500\text{ m} \qquad (2)$$

Physically, this area is equal to the *net displacement* Δx of the particle during the time interval from $t = 0$ to $t = 10$ s.

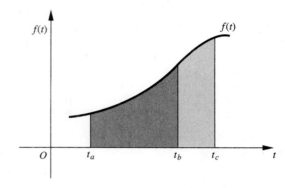

FIGURE 6.2 The area under the curve from t_a to t_c is equal to the sum of the definite integrals of $f(t)$ from t_a to t_b and t_b to t_c (see Equation 6.10).

Definite integrals are *additive*. In Figure 6.2 it is clear that the area under the curve from t_a to t_c is composed of two parts, and that the total area is the sum of the area from t_a to t_b and the area from t_b to t_c. In terms of definite integrals, this statement is expressed as

$$\int_{t_a}^{t_b} f(t)\, dt + \int_{t_b}^{t_c} f(t)\, dt = \int_{t_a}^{t_c} f(t)\, dt \tag{6.10}$$

We can generalize this result to any ordering of t_a, t_b, and t_c provided we define

$$\int_{t_a}^{t_b} f(t)\, dt = -\int_{t_b}^{t_a} f(t)\, dt \tag{6.11}$$

This equation states that the definite integral carried for t changing in one direction (t_a to t_b) is equal to the *negative* of the same definite integral carried out for t changing in the opposite direction (t_b to t_a).

It should also be pointed out that definite integrals possess the property of linearity, as do indefinite integrals (Equation 6.4).

In the examples discussed earlier in this section, the functions $f(t)$ have been relatively simple, and the definite integrals were evaluated by elementary geometric arguments. But how do we evaluate a definite integral if the integrand is a complicated function? The general procedure is illustrated in Figure 6.3. Here we need to determine the area under a complicated curve between two limits, t_a and t_b. We divide this interval into a number n of subintervals with widths Δt_1, Δt_2, $\ldots$, Δt_n. For simplicity we take the width of each subinterval to be the same. Because the total width is $t_b - t_a$ and there are n subintervals, the width of each subinterval is

$$\Delta t = \frac{1}{n} (t_b - t_a) \tag{6.12}$$

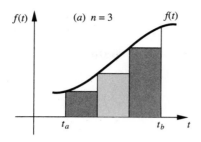 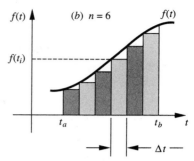

FIGURE 6.3 Determining the area under a curve by summing the areas of a number of rectangles that fit under the curve. The error in this procedure (the unshaded areas above the rectangles) diminishes as n becomes large, and when n tends to infinity, the definite integral is exactly equal to the sum of the rectangular areas.

In Figure 6.3 the left-hand edges of the subintervals are located at

$$t_1 = t_a$$
$$t_2 = t_1 + \Delta t$$
$$t_3 = t_2 + \Delta t$$

and so forth. The area of the ith rectangle is equal to $f(t_i)\Delta t$, and the *total* area of all of the n rectangles between t_a and t_b can be expressed as the sum,

$$A_n = \sum_{i=1}^{n} f(t_i)\Delta t \tag{6.13}$$

Comparing Figures 6.3a and 6.3b, it is clear that the *larger* the value of n, the closer A_n approximates the *true* area under the curve. In fact, for sufficiently "smooth" functions $f(t)$, as n tends to infinity (and Δt tends to zero), A_n becomes exactly equal to the true area under the curve between t_a and t_b. Combining this result with Equation (6.8), we find

$$D = \int_{t_a}^{t_b} f(t)\, dt = \lim_{n \to \infty} \sum_{i=1}^{n} f(t_i)\Delta t \tag{6.14}$$

which is a fundamental result of integral calculus.

The reason for the notation used for the definite integral is now apparent. Symbolically, dt replaces Δt, the integral sign $\int$ replaces the summation sign Σ, and the limits of integration, t_a and t_b, replace the limits of summation 1 and n.

From a practical standpoint it is important to note that integration on a computer is often performed by evaluating the summation in Equation (6.13) for large but finite n. The value of n is adjusted to give the minimum error compatible with available computer time (and memory) and the desired accuracy.

EXERCISES   *Evaluate $D = \int_{t_a}^{t_b} v(t)\, dt$ for the following cases:*

1.

(Ans. 256)

3.

(Ans. 222)

2.

(Ans. 1)

4.

(Ans. 341)

Evaluate the definite integral $D = \int_{t_a}^{t_b} v(t)\, dt$ for the following special cases. (State the units in each case.)

5. In Exercise 4 above, $t_b = 10^7$ s and $v_0 = \pi$ m/s.   (Ans. 19)

6. In Exercise 1 above, $t_b = 10$ μs and $v_0 = 3 \times 10^6$ m/s.   (Ans. 358)

7. In Exercise 2 above, $t_b = 2$ s and $v_0 = 5$ m/s.   (Ans. 189)

8. In Exercise 3 above, $t_b = 3$ minutes, and $v_0 = 60$ m/s.   (Ans. 271)

9. What is the physical significance of D in the above exercises?   (Ans. 297)

Evaluate $D = \int_{x_a}^{x_b} f(x)\, dx$ in the following cases (although the variable is taken to be x, all of the results that have been developed in Section 6.2 are valid):

10. $f(x) = 6$, $x_a = 0$, and $x_b = 2$ m   (Ans. 13)

11. $f(x) = 6$, $x_a = 2$ m, and $x_b = 0$ (Ans. 321)

12. $f(x) = -5$, $x_a = 0$, and $x_b = 3$ m   (Ans. 185)

13. $f(x) = 3x$ (in cm), $x_a = 0$, and $x_b = 15$ cm   (Ans. 365)

14. $f(x) = 6 + 3x$ (in cm), $x_a = 0$, and $x_b = 15$ cm   (Ans. 379)

6.3 FUNDAMENTAL THEOREM OF INTEGRAL CALCULUS

In this section we derive the simple but very important relation between definite and indefinite integrals. If the upper limit of the definite integral in Equation (6.8) is a *variable t* rather than a *fixed number* t_b, then D is a *variable function* of t. The notation for the definite integral with a variable upper limit is:

$$D(t) = \int_{t_a}^{t} f(t')\, dt' \qquad \text{(6.15)}$$

To avoid confusion between the *upper limit* and the *integration variable,* t has been replaced by t' in the integrand of Equation (6.15).

How does $D(t)$ vary with t? Let us compare $D(t + \Delta t)$ with $D(t)$ for small increments Δt. Referring to Figure 6.4 and Equation (6.10) (with $t_b = t$ and $t_c = t + \Delta t$), it follows that

$$\int_{t_a}^{t} f(t')\, dt' + \int_{t}^{t+\Delta t} f(t')\, dt' = \int_{t_a}^{t+\Delta t} f(t')\, dt'$$

Comparing this equation with Equation (6.15), we find

$$D(t) + \int_{t}^{t+\Delta t} f(t')\, dt' = D(t + \Delta t) \qquad \text{(6.16)}$$

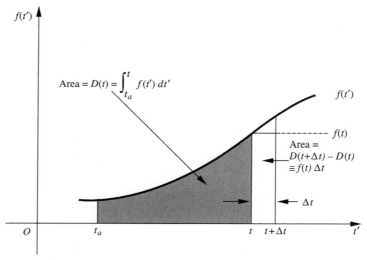

FIGURE 6.4 Graphical demonstration that $D(t + \Delta t) - D(t) \cong f(t)\Delta t$ for small increments Δt (see Equation 6.17).

From Figure 6.4 it is clear that the area under the graph of $f(t')$, between the limits t and $t + \Delta t$, can be approximated by

$$\int_t^{t+\Delta t} f(t')\, dt' \cong f(t)\Delta t$$

for small values of Δt. Combining this result with Equation (6.16), and rearranging terms, we find

$$D(t + \Delta t) - D(t) \cong f(t)\Delta t \tag{6.17}$$

Dividing by Δt gives

$$\frac{D(t + \Delta t) - D(t)}{\Delta t} \cong f(t) \tag{6.18}$$

In the limit $\Delta t \to 0$, the approximate equality becomes exact, and the left-hand side of Equation (6.18) becomes the derivative of $D(t)$ with respect to t (see Section 4.2). That is,

$$\frac{dD}{dt} = \lim_{\Delta t \to 0} \frac{D(t + \Delta t) - D(t)}{\Delta t} = f(t) \tag{6.19}$$

Combining Equations (6.1) and (6.19), we find that the *indefinite integral* $I(t)$ and *definite integral* $D(t)$ are related by

$$\frac{dI}{dt} = f(t) = \frac{dD}{dt} \tag{6.20}$$

Since $I(t)$ and $D(t)$ have the same derivative, $f(t)$, they differ only by a constant. Therefore, we can write

$$D(t) = \int_{t_a}^{t} f(t')\, dt' = I(t) + C \tag{6.21}$$

where C is a constant. To determine the value of C, we take $t = t_a$ in Equation (6.21). Because the upper and lower limits of the integral are then the same, the integral is zero, and we have

$$0 = I(t_a) + C$$

Therefore,

$$C = -I(t_a)$$

and Equation (6.21) can be expressed as

$$D(t) = \int_{t_a}^{t} f(t')\, dt' = I(t) - I(t_a)$$

(6.22)

Equation (6.22), which is a precise relation between the definite integral $D(t)$ and the indefinite integral $I(t)$, is known as the *fundamental theorem of integral calculus.*

Because $I(t)$ is known for a broad class of functions $f(t)$ (see, for example, Table 6.1), Equation (6.22) provides a valuable method for evaluating $D(t)$ for situations in which the area under the $f(t)$ *versus* t curve cannot be calculated by simple geometric arguments.

▶ **EXAMPLE 6.3.1** Evaluate the definite integral

$$D(t) = \int_{0}^{t} \alpha t'^{6}\, dt'$$

(1)

where α is a constant.

Comparing (1) with Equation (6.22) gives

$$D(t) = \int_{0}^{t} \alpha t'^{6}\, dt' = I(t) - I(0)$$

(2)

where $I(t)$ is the indefinite integral of $f(t) = \alpha t^{6}$. Referring to Table 6.1, we find

$$I(t) = \alpha\left(\frac{1}{7} t^{7} + C\right)$$

(3)

where C is an arbitrary constant. Therefore,

$$D(t) = \int_{0}^{t} \alpha t'^{6}\, dt' = \alpha\left(\frac{1}{7} t^{7} + C\right) - \alpha(0 + C)$$

$$= \frac{\alpha}{7} t^{7}$$

(4)

Note that the arbitrary constant C does not occur in the final expression for $D(t)$. In fact, whenever a *definite* integral is evaluated, the constant will cancel; therefore, we can dispense with writing down the constant in such situations.

It is customary to write Equation (6.22) in the form

$$D(t) = \int_{t_a}^{t} f(t') \, dt' = I(t) \Big|_{t_a}^{t} \tag{6.23}$$

where

$$I(t) \Big|_{t_a}^{t} \quad \text{denotes } I(t) - I(t_a)$$

If the upper limit t_b is reinstated in Equation (6.23), then the definite integral between *fixed* limits is (compare Equation 6.8)

$$D = \int_{t_a}^{t_b} f(t) \, dt = I(t) \Big|_{t_a}^{t_b} \tag{6.24}$$

It is important to note that Equations (6.22)–(6.24), which are *all* forms of the fundamental theorem of integral calculus, provide the *easiest* method for evaluating definite integrals.

▶ **EXAMPLE 6.3.2**   Evaluate the definite integral

$$D = \int_{4}^{6} t^3 \, dt \tag{1}$$

Comparing (1) with Equation (6.24) gives

$$\int_{4}^{6} t^3 \, dt = I(t) \Big|_{4}^{6} \tag{2}$$

where $I(t)$ is the indefinite integral of $f(t) = t^3$. From Table 6.1, we find (dispensing with the integration constant C because we are evaluating a definite integral)

$$I(t) = \frac{t^4}{4} \tag{3}$$

Substituting (3) into (2) gives

$$\int_{4}^{6} t^3 \, dt = \frac{1}{4} t^4 \Big|_{4}^{6} = \frac{1}{4} (6^4 - 4^4)$$

$$= \frac{1}{4} (1296 - 256) = 260 \tag{4}$$

▶ **EXAMPLE 6.3.3**   Evaluate

$$D = \int_0^{\pi/2} \cos t \, dt \tag{1}$$

Using Table 6.1 to find the integral of cos t, we can write

$$D = \sin t \Big|_0^{\pi/2} = \sin \pi/2 - \sin 0$$

$$= 1 - 0 \tag{2}$$

Therefore,

$$\int_0^{\pi/2} \cos t \, dt = 1 \tag{3}$$

Look back again over Examples 6.3.1, 6.3.2, and 6.3.3 and notice how the procedure has been successively simplified. In Example 6.3.1, we first obtained the result in terms of indefinite integrals, then we evaluated each, carrying the constant of integration, which finally cancelled. In Example 6.3.2, we again found the indefinite integral but dispensed with exhibiting the constant. Finally, in Example 6.3.3, we immediately wrote down the integral from Table 6.1, showing the evaluation at the end points; that is, we used Equation (6.24) directly. This latter procedure is much simpler and we will follow it henceforth. But it is well to remember how we arrived at this point.

▶ **EXAMPLE 6.3.4**   When an object falls through air, it does not continue to accelerate indefinitely. Frictional effects tend to retard (or damp) the motion. This situation can be approximated by writing the net force acting on the object as a decreasing exponential. That is,

$$F(t) = ma_0 e^{-t/\tau} \tag{1}$$

where m, a_0, and τ are constants. If the object is dropped from rest at $t = 0$, the initial condition is $v(t = 0) = 0$. What is the velocity $v(t)$ at later times?

First, we express the acceleration as

$$a(t) = \frac{F(t)}{m} = a_0 e^{-t/\tau} \tag{2}$$

Next, we write

$$v(t) = \int_0^t a(t')\,dt'$$

$$= a_0 \int_0^t e^{-t'/\tau}\,dt' \tag{3}$$

Using the appropriate entry in Table 6.1, we have

$$v(t) = a_0(-\tau e^{-t'/\tau}) \Big|_0^t \tag{4}$$

Because $e^0 = 1$, we obtain, finally,

$$v(t) = a_0\tau(1 - e^{-t/\tau}) \tag{5}$$

Looking at (2) and (5), we see that the effect of air resistance is to cause the acceleration of the object gradually to approach zero. (As t becomes large compared to τ, $e^{-t/\tau} \to 0$.) Moreover, (5) shows that as $a(t) \to 0$, the velocity $v(t)$ tends toward the constant velocity $a_0\tau$, which we call the *terminal velocity*.

EXERCISES *Evaluate $D(t) = \int_0^t f(t')\,dt'$ for the following choices of $f(t)$:*

1. $f(t) = t^6$ (Ans. 369) **4.** $f(t) = t^{-n}$, for $n \leq 0$ (Ans. 350)
2. $f(t) = 5e^{-at}$, where a = constant (Ans. 190) **5.** $f(t) = 2\ln(t+1)$ (Ans. 90)
3. $f(t) = a\sin 6t$, where a = constant (Ans. 46)

A particle of mass m moves along the X-axis under the influence of a variable force F(t). Its initial velocity is $v(t = 0) = 0$. Determine the particle's velocity $v(t)$ at an arbitrary time t for the following choices of $F(t)$:

6. $F(t) = 8$ (Ans. 346) **8.** $F(t) = at^{3/2}$, where a = constant (Ans. 221)
7. $F(t) = F_0\cos 2t$, where F_0 = constant (Ans. 364) **9.** $F(t) = F_0 e^{-t/\tau}$, where F_0 and τ are constants
(Ans. 290)

Evaluate $D = \int_{x_a}^{x_b} f(x)\,dx$ for the following choices of $f(x)$, x_a and x_b:

10. $f(x) = x + 2x^2$, for $x_a = 0$ and $x_b = 2$ (Ans. 48) **12.** $f(x) = e^{2x}$, for $x_a = 1$ and $x_b = 2$ (Ans. 205)
11. $f(x) = \cos \pi x$, for $x_a = 0$ and $x_b = 1$ (Ans. 381) **13.** $f(x) = \sin 2x$, for $x_a = 0$ and $x_b = \pi/2$ (Ans. 375)

6.4   WORK AS AN INTEGRAL

In Section 5.6 we discussed the fact that the *work* done by a force is equal to the product of the (constant) force and the distance through which the force acts. If an

FIGURE 6.5 A mass m is moved from $x = x_a$ to $x = x_b$ by the variable force $F(x)$.

object experiences a force $F(x)$ that varies with position x, then the work done by the force in displacing the object from one location to another can be expressed as an integral.

Consider the one-dimensional motion illustrated in Figure 6.5. An object of mass m moves along the X-axis as the result of a variable force $F(x)$. How much work is done by the force $F(x)$ in displacing the object from $x = x_a$ to $x = x_b$?

First consider the case in which $F(x) = F_0 = $ constant. From the discussion in Section 5.6, it follows that

$$\text{Work done by the force } F_0 = (\text{Force}) \times (\text{Net displacement})$$
$$= F_0(x_b - x_a)$$

If we denote the work done in displacing the object from $x = x_a$ to $x = x_b$ by W_{ab}, then

$$W_{ab} = F_0(x_b - x_a) \tag{6.25}$$

The right-hand side of Equation (6.25) is equal to the definite integral of the constant function F_0 between the limits x_a and x_b (see Sections 6.2 and 6.3):

$$F_0(x_b - x_a) = \int_{x_a}^{x_b} F_0 \, dx$$

Combining this result with Equation (6.25) gives

$$W_{ab} = \int_{x_a}^{x_b} F_0 \, dx \tag{6.26}$$

Graphically, the work done is equal to the area under the curve $F(x) = F_0 = $ constant, between the limits $x = x_a$ and $x = x_b$ (see Fig. 6.6).

FIGURE 6.6 Work done by the constant force F_0 is equal to the shaded area.

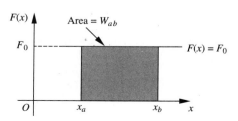

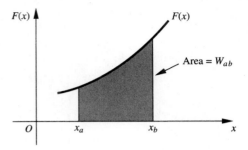

FIGURE 6.7 Work done by the variable force $F(x)$ is equal to the shaded area.

The form of Equation (6.26) is actually quite general. If the object experiences a *variable* force $F(x)$ and is displaced from $x = x_a$ to $x = x_b$ (see Fig. 6.7), then the work done by the force $F(x)$ can be expressed as the definite integral

$$W_{ab} = \int_{x_a}^{x_b} F(x)\, dx \tag{6.27}$$

Graphically, the work done is equal to the area under the curve $F(x)$, between the limits $x = x_a$ and $x = x_b$ (see Fig. 6.7).

▶ **EXAMPLE 6.4.1**   As illustrated in the accompanying diagram, a mass m rests on a frictionless surface and is attached to a spring, with spring constant k.

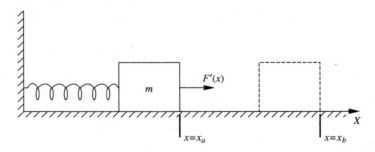

(a)   What force $F'(x)$ is required to displace the mass *at constant velocity?*
(b)   What is the work done by $F'(x)$ in displacing the mass from $x = 0$ to $x = x_b$?

In answer to part (*a*), the spring exerts a *restoring* force $F(x) = -kx$ on the mass. In order to displace the mass at constant velocity it is necessary that the externally applied force $F'(x)$ be *equal in magnitude and opposite in direction* to $F(x)$. Therefore,

$$F'(x) = kx \tag{1}$$

In answer to part (b), the work done by the force $F'(x)$ in displacing the mass from $x = 0$ to $x = x_b$ is (see Equation 6.27 with $x_a = 0$)

$$W_{ab} = \int_0^{x_b} F'(x)\, dx = \int_0^{x_b} kx\, dx \tag{2}$$

Evaluating the integral, we find

$$W_{ab} = \int_0^{x_b} kx\, dx = \frac{1}{2} kx^2 \bigg|_0^{x_b} = \frac{1}{2} k\, (x_b^2 - 0) \tag{3}$$

Therefore the work done by $F'(x) = kx$ in displacing the mass from $x = 0$ to $x = x_b$ is

$$W_{ab} = \frac{1}{2} kx_b^2 \tag{4}$$

As illustrated in the diagram below, the result expressed in (4) can also be obtained graphically by computing the area of the shaded triangle under the curve $F'(x) = kx$, between the limits $x = 0$ and $x = x_b$. (What is the area?)

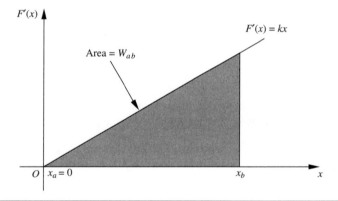

Work can also be expressed as an integral in two and three dimensions, but we will consider only the two-dimensional case. Suppose that a particle moves in the X-Y plane along the trajectory indicated by the solid curve in Figure 6.8. Assume also that the particle experiences a force,

$$\mathbf{F} = F_x \mathbf{i} + F_y \mathbf{j} \tag{6.28}$$

where $\mathbf{i}$ and $\mathbf{j}$ are unit vectors in the X- and Y-directions, respectively. In general, the force components, F_x and F_y, are variable and depend on the position of the particle which is specified by the displacement vector,

$$\mathbf{s} = x \mathbf{i} + y \mathbf{j} \tag{6.29}$$

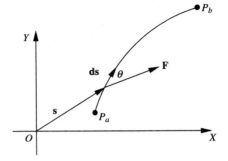

FIGURE 6.8   A particle moves along the curve from P_a to P_b under the influence of the variable force **F**. The quantity **ds** represents an infinitesimal displacement *along* the curve.

We have already seen (in Section 5.6) that the work done by a constant force is **F** · **s**. If the displacement is an infinitesimal distance, **ds**, then the incremental amount of work done is **F** · **ds**, where

$$\mathbf{ds} = \mathbf{i}\,dx + \mathbf{j}\,dy \tag{6.30}$$

Furthermore, if the force **F** changes with position, the total work done in moving the particle from one position P_a to another P_b along a particular path must be computed from an integral,

$$W_{ab} = \int_{P_a}^{P_b} \mathbf{F} \cdot \mathbf{ds} \tag{6.31}$$

Because the particle moves along a particular path, the integral of **F** · **ds** must be computed along this same path. Such an integral is called a *line integral.* In order to make this type of calculation, we first use Equations (6.28) and (6.30) to compute **F** · **ds**:

$$\mathbf{F} \cdot \mathbf{ds} = (F_x\,\mathbf{i} + F_y\,\mathbf{j}) \cdot (\mathbf{i}\,dx + \mathbf{j}\,dy)$$

Making use of Equations (5.32) for $\mathbf{i} \cdot \mathbf{i}$, $\mathbf{i} \cdot \mathbf{j}$, and so forth, we obtain

$$\mathbf{F} \cdot \mathbf{ds} = F_x\,dx + F_y\,dy \tag{6.32}$$

Therefore, the evaluation of the integral in Equation (6.31) separates into two parts,

$$W_{ab} = \int_{P_a}^{P_b} F_x\,dx + \int_{P_a}^{P_b} F_y\,dy \tag{6.33}$$

where both integrals must be evaluated along the particular path from P_a to P_b.

▶   **EXAMPLE 6.4.2**   Suppose that we move a particle of mass m at constant velocity against the force of gravity from P_a to P_b along the dotted path in the figure below. How much work has been done?

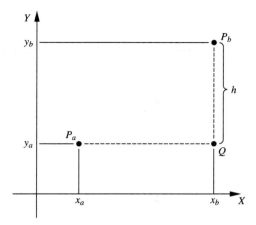

The gravitational force is directed downward and has a magnitude mg. Therefore, in order to move the particle at constant velocity, we must exert an *upward* force of this same magnitude. This force is

$$\mathbf{F} = mg\,\mathbf{j} \tag{1}$$

Notice that there is no X-component of $\mathbf{F}$. (Why?) Thus,

$$F_x = 0; \qquad F_y = mg \tag{2}$$

and the work integral becomes

$$W_{ab} = \int_{P_a}^{P_b} F_y\,dy = \int_{P_a}^{P_b} mg\,dy \tag{3}$$

where the path is specified by the dotted line in the figure. We can further decompose the integral by considering separately the two parts of the path.

$$W_{ab} = \int_{P_a}^{Q} mg\,dy + \int_{Q}^{P_b} mg\,dy \tag{4}$$

Now, along the path $P_a \rightarrow Q$, the Y-coordinate does not change, and any integral involving dy will equal zero. We are therefore left with

$$W_{ab} = \int_{Q}^{P_b} mg\,dy \tag{5}$$

But the point Q is (x_b, y_a) and the point P_b is (x_b, y_b), and the path is a straight line from Q to P_b. Thus, (5) simplifies to

$$W_{ab} = \int_{y_a}^{y_b} mg\,dy = mg(y_b - y_a) \tag{6}$$

Finally, we note in the figure that $y_b - y_a = h$, so our result is

$$W_{ab} = mgh \tag{7}$$

which is the familiar expression for the work required to raise an object by a distance h near the surface of the earth.

It is interesting to note that in this case (as in any case involving a constant force), the work required to displace an object through a certain distance depends only on how far the object is moved along the line of the force. Some other possible paths connecting P_a and P_b are examined in the exercises.

▶ **EXAMPLE 6.4.3**   A mass m undergoes *two-dimensional* simple harmonic motion in the X-Y plane with *restoring* force components

$$
\begin{aligned}
F_x &= -kx \\
F_y &= -ky
\end{aligned}
\tag{1}
$$

Paralleling the analysis in Example 6.4.1, the *externally imposed* force components required to displace the mass with constant velocity can be expressed as

$$
\begin{aligned}
F_x' &= kx \\
F_y' &= ky
\end{aligned}
\tag{2}
$$

What is the work done by the force

$$F' = F_x' \, i + F_y' \, j \tag{3}$$

in displacing the mass from the point P_a at (x_a, y_a) to the point P_d at (x_d, y_d) along the following paths in the figure:

(a) *Horizontally* from P_a at (x_a, y_a) to P_b at (x_d, y_a), and *vertically* from P_b at (x_d, y_a) to P_d at (x_d, y_d)?
(b) *Vertically* from P_a at (x_a, y_a) to P_c at (x_a, y_d), and *horizontally* from P_c at (x_a, y_d) to P_d at (x_d, y_d)?
(c) Directly *along the diagonal* in the figure from P_a at (x_a, y_a) to P_d at (x_d, y_d)?

In answer to part (a), $dy = 0$ along the path from P_a to P_b. Therefore, the work integral from P_a to P_b becomes

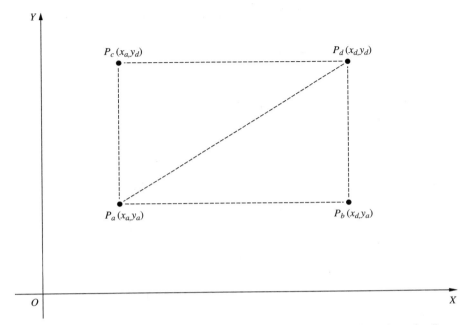

FIGURE 6.4.3   A mass m experiences the force in (2) and is displaced from the point P_a to the point P_d along three possible paths: (a) P_a to P_b to P_d, (b) P_a to P_c to P_d, and (c) directly along the diagonal from P_a to P_d.

$$W_{ab} = \int_{P_a}^{P_b} F'_x \, dx$$

$$= \int_{x_a}^{x_d} kx \, dx$$

$$= \frac{1}{2} k(x_d^2 - x_a^2) \tag{4}$$

Similarly, $dx = 0$ along the path from P_b to P_d, and the work integral is

$$W_{bd} = \int_{P_b}^{P_d} F'_y \, dy$$

$$= \int_{y_a}^{y_d} ky \, dy$$

$$= \frac{1}{2} k(y_d^2 - y_a^2) \tag{5}$$

Therefore, the *total* work along the path from P_a to P_b to P_d is

$$W_{ab} + W_{bd} = \frac{1}{2}k(x_d^2 - x_a^2) + \frac{1}{2}k(y_d^2 - y_a^2) \tag{6}$$

In answer to part (b), $dx = 0$ along the path from P_a to P_c, and the work integral is

$$\begin{aligned}
W_{ac} &= \int_{P_a}^{P_c} F_y'\, dy \\
&= \int_{y_a}^{y_d} ky\, dy \\
&= \frac{1}{2}k(y_d^2 - y_a^2)
\end{aligned} \tag{7}$$

Similarly, $dy = 0$ along the path from P_c to P_d, and the work integral is

$$\begin{aligned}
W_{cd} &= \int_{P_c}^{P_d} F_x'\, dx \\
&= \int_{x_a}^{x_d} kx\, dx \\
&= \frac{1}{2}k(x_d^2 - x_a^2)
\end{aligned} \tag{8}$$

Therefore, the *total* work along the path from P_a to P_c to P_d is

$$W_{ac} + W_{cd} = \frac{1}{2}k(y_d^2 - y_a^2) + \frac{1}{2}k(x_d^2 - x_a^2) \tag{9}$$

which is identical to the result obtained in Equation (6) by taking the path from P_a to P_b to P_d.

In answer to part (c), the work integral along the *diagonal* path from P_a to P_d is

$$\begin{aligned}
W_{ad} &= \int_{P_a}^{P_d} F_x'\, dx + \int_{P_a}^{P_d} F_y'\, dy \\
&= \int_{x_a}^{x_d} kx\, dx + \int_{y_a}^{y_d} ky\, dy \\
&= \frac{1}{2}k(x_d^2 - x_a^2) + \frac{1}{2}k(y_d^2 - y_a^2)
\end{aligned} \tag{10}$$

We note from Equations (6), (9), and (10) that

$$W_{ab} + W_{bd} = W_{ac} + W_{cd} = W_{ad} \tag{11}$$

That is, the work in moving the mass m from the point P_a to the point P_d is the *same* in all three cases. This feature that the work is *independent of the path taken* is characteristic of *conservative force fields,* of which Equation (1) is an example.

EXERCISES   *A mass is displaced from $x = 0$ to $x = x_b$ by a variable force $F(x)$. Calculate the work done for the following choices of $F(x)$:*

1. $F(x) = kx^3$, where k is a constant   (Ans. 80)

2. $F(x) = F_0 \sin \alpha x$, where α and F_0 are constants

(Ans. 127)

3. $F(x) = F_0 e^{\alpha x}$, where α and F_0 are constants

(Ans. 14)

4. $F(x) = kx^3 - \alpha x$, where k and α are constants

(Ans. 376)

Rework Example 6.4.2 for the cases in which the particle is moved along the following paths:

5. $P_a \rightarrow (x_a, y_b) \rightarrow P_b$.   (Ans. 325)

6. $P_a \rightarrow (x_b + a, y_a) \rightarrow (x_b + a, y_b) \rightarrow P_b$.   (Ans. 82)

7. $P_a \rightarrow (x_a, y_b + h) \rightarrow (x_b, y_b + h) \rightarrow P_b$. (Ans. 230)

7

Experimental Error and Significant Figures

7.1   EXPERIMENTAL NUMBERS

In physics we seek to provide an orderly and precise description of natural phenomena through the use of *numbers*. In the report of an experiment, the use of numbers instead of qualitative descriptions is always preferred. For example, suppose that we examine the light emitted in a certain physical process. We could report the result of the experiment by saying "The color of the light is yellow." But we give a much more precise statement of the result if we say "The emitted light has a wavelength of 5.236×10^{-7} m."

More is involved in reporting the result of an experiment than a simple statement of the experimental number. To state that "$\lambda = 5.236 \times 10^{-7}$ m" leaves unanswered the question, "How precise is the result?" Could the wavelength actually be 5.300×10^{-7} m? An experimenter must always carefully analyze results in terms of the measuring standards used, the reliability of the equipment, and any factors that could influence the result. After this is done and after the quantity has been measured many times to check the reproducibility of the instruments and technique, the experimenter will be in a position to state the result in the following form:

$$\lambda = (5.236 \pm 0.002) \times 10^{-7} \text{ m}$$

The quantity $\pm 0.002 \times 10^{-7}$ m that is attached to the value $\lambda = 5.236 \times 10^{-7}$ m is the experimenter's way of indicating the precision of the result. We call this quantity the "uncertainty" in the result and we label it $\delta\lambda$. The value of 5.236×10^{-7} m is the *average* or *mean* of all the measurements; we denote this value by $\bar{\lambda}$. Thus, the report of the experiment is in the form,

TABLE 7.1   Wavelength Data

i	λ_i ($\times\ 10^{-7}$ m)	$\|\Delta_i\| = \|\overline{\lambda} - \lambda_i\|$ ($\times\ 10^{-10}$ m)
1	5.240	4
2	5.237	1
3	5.231	5
4	5.236	0
5	5.238	2
6	5.234	2
7	5.236	0
8	5.236	0
9	5.241	5
10	5.235	1
11	5.232	4
12	5.236	0

$$\lambda = \overline{\lambda} \pm \delta\lambda \tag{7.1}$$

What is the meaning of a result stated in this way? What is the significance of the range of wavelength values that extends from $\overline{\lambda} - \delta\lambda$ to $\overline{\lambda} + \delta\lambda$? Clearly, the experimenter believes that the ''true'' value of λ lies in this range and, to the best ability to measure the wavelength, the experimenter believes that the *most probable* value of λ is $\overline{\lambda}$. If another measurement of the same wavelength is measured by a different experimenter with the result $\lambda = (5.2368 \pm 0.0007) \times 10^{-7}$ m, we would say that the two values are in agreement, because the new value (which is more precise) lies within the range of values assigned by the first experimenter. In obtaining a result to which is attached a smaller value of $\delta\lambda$, the second experimenter has improved our knowledge of this particular wavelength.

Let us examine the data of the first experimenter and see how this experimenter arrived at this result. The values of the 12 independent measurements are listed in Table 7.1. The mean value $\overline{\lambda}$ is calculated from

$$\overline{\lambda} = \frac{1}{N}\sum_{i=1}^{N} \lambda_i \tag{7.2}$$

where N is the total number of values (here, 12). Adding the individual values of the wavelength and dividing by 12 gives $\overline{\lambda} = 5.236 \times 10^{-7}$ m.

Having determined $\overline{\lambda}$, we now calculate the difference between $\overline{\lambda}$ and each individual measurement, λ_i. We call this difference the *deviation from the mean:*

$$\Delta_i = \overline{\lambda} - \lambda_i \tag{7.3}$$

The average of the Δ_i is clearly *zero* (this is, in fact, the definition of the mean value). Therefore, as a measure of the reproducibility of the experimental values, we use the *magnitudes* of the individual deviations, $|\Delta_i|$, and calculate the average:

$$\delta\lambda = \frac{1}{N}\sum_{i=1}^{N}|\Delta_i| \tag{7.4}$$

This quantity is called the *average deviation*. The $|\Delta_i|$ are listed in Table 7.1. (Notice that the units for the $|\Delta_i|$ are smaller than those for the λ_i, $|\Delta_i| = 4$ means $|\overline{\lambda} - \lambda_i| = 0.004 \times 10^{-7}$ m.) For the data tabulated, we find $\delta\lambda = 0.002 \times 10^{-7}$ m. Therefore, the result of the experiment is stated as $\lambda = (5.236 \pm 0.002) \times 10^{-7}$ m. The quantity $\delta\lambda$ is often simply called the "experimental error."

The procedure outlined here—namely, the assignment of the average deviation as the uncertainty in the result—is probably the most satisfactory one to use when treating experimental data obtained in the physics laboratory. A procedure based on statistical arguments leads to a different method for stating the experimental uncertainty, but this procedure supposes a very large sample of data. Because relatively small samples of data are usually taken in most laboratory courses, the average deviation tends to be a more meaningful expression of the precision. (One point to bear in mind, however, is that when comparing experimental uncertainties, be sure that they were computed using the same procedure!)

(If a very large sample of data is available, the statistically proper method for computing the uncertainty is to use the expression for the *standard deviation* [S.D.]:

$$\text{S.D.} = 0.6745\sqrt{\frac{\sum_{i=1}^{N}|\Delta_i|^2}{N(N-1)}}$$

When a result is expressed as $\lambda = \overline{\lambda} \pm \text{S.D.}$, it means that there is a probability of 0.5 that the "true" value of λ lies within the range from $\overline{\lambda} - \text{S.D.}$ to $\overline{\lambda} + \text{S.D.}$ Applying this procedure to the data of Table 7.1 yields S.D. $= 0.0006 \times 10^{-7}$ m, a much smaller uncertainty than that obtained by computing the *average deviation*. But a standard deviation cannot meaningfully be applied to a set of only 12 data.)

It must be emphasized again that the choice of the average deviation as an indicator of the precision of the experiment is *completely arbitrary,* and some persons will prefer to use only the standard deviation. However, statistical methods really apply only to large samples of data and their use with small samples will frequently give a gross overestimate of the reliability of the experiment. Someone will always be able to interpret properly your results if you specify the procedure used to reduce your data.

EXERCISES

1. The following are "experimental" values of π, determined by several individuals who measured the circumference and the diameter of a circle and calculated the ratio: 3.141 ± 0.001, 3.144 ± 0.002, $3.140 \pm$

0.002, 3.15 ± 0.01, 3.1416 ± 0.0005. Which of the results is in agreement with the true value? (Notice that the measurements here are of a *mathematical* quantity, not a *physical* quantity, so a "true" value does exist.) If you were to choose one of the values as the most reliable, which would it be?   (Ans. 49)

2. Calculate the mean value and the average deviation for the following set of measurements of the acceleration due to gravity. (All values are in units of m/s^2.)
g = 9.6, 9.9, 10.0, 9.2, 10.2, 9.8, 9.7, 10.0, 10.3, 9.3.
(Ans. 374)

7.2   ABSOLUTE AND RELATIVE ERROR

If the result of a certain length measurement is stated as L = 3.08 ± 0.03 cm, we mean that the *absolute error* in L is 0.03 cm. (Actually, we properly use the term *absolute uncertainty* or *absolute average deviation,* but "absolute error" is more commonly used.) The error is "absolute" in the sense that it is stated in the same physical units as the quantity measured. The *relative error* in L is the fraction of L represented by the absolute error. Let the absolute error in L be δL; then, the specification of a measurement of L is given by a statement of the form $L \pm \delta L$. Thus,

$$\text{Relative error in } L = \frac{\delta L}{L} \tag{7.5}$$

For example, if $L \pm \delta L$ = 3.08 ± 0.03 cm, then the absolute and relative errors are

$$\text{Absolute error} = \delta L = 0.03 \text{ cm}$$

$$\text{Relative error} = \frac{\delta L}{L} = \frac{0.03 \text{ cm}}{3.08 \text{ cm}} = 0.01 \ (1\%)$$

That is, the relative error in L is 0.01 or 1% of the value of L. The smaller the relative error, the more precise is the measurement. But a smaller *absolute* error does not necessarily mean an improvement in the relative error. For example, the result L' = 1.07 ± 0.02 cm has a smaller absolute error than that assigned to L above, but the relative error in L' is 2% and so this measurement is less precise in a relative sense.

Very precise measurements sometimes carry relative errors that are given in *parts per million* (ppm). A relative error of 1 ppm is 0.000 001 ($10^{-4}\%$).

Occasionally one sees a result stated as $(3.98 \times 10^4 \text{ cm}) \pm 2\%$. This means that the relative error is 0.02, and if the absolute error is desired, it must be calculated: in this case, absolute error = $0.02 \times (3.98 \times 10^4 \text{ cm}) = 0.08 \times 10^4 \text{ cm} = 800 \text{ cm}$.

▶ **EXAMPLE 7.2.1**   What is the precision of our present knowledge of the velocity of light?

According to a recent careful study of all the various measurements of the velocity of light, the value is

$$c = (2.997\ 925 \pm 0.000\ 001) \times 10^8 \text{ m/s}$$

$$\text{Absolute error} = 0.000\ 001 \times 10^8 \text{ m/s} = 100 \text{ m/s}$$

$$\text{Relative error} = \frac{0.000\ 001 \times 10^8 \text{ m/s}}{2.997\ 925 \times 10^8 \text{ m/s}} = 0.000\ 000\ 33$$

$$= 0.33 \text{ ppm}$$

The velocity of light is the most precisely known of all the fundamental physical constants.

EXERCISES   *What is the relative error in each of the following values:*

1. 16.45 ± 0.03 m   (Ans. 208)   **3.** 563 ± 2 m/s   (Ans. 213)
2. 1.234 ± 0.001 N   (Ans. 220)

What is the relative error (in ppm) in each of the following values:

4. $3.141592 \pm 0.000002 \text{ s}^{-1}$ (Ans. 54)   **6.** 939.553 ± 0.003 kg   (Ans. 269)
5. 26.9815 ± 0.0001 m   (Ans. 131)

7.3 SIGNIFICANT FIGURES

When we refer to an experimental result, we often do not give explicitly the error. That is, instead of writing $L = 3.264 \pm 0.002$ m, we write only $L = 3.264$ m. We imply by this shorthand notation that the uncertainty in the result is in the *last digit* given (but we do not know whether the uncertainty is ± 1 digit, ± 2 digits, or ± 5 digits). Thus, if we write $L = 3.264$ m, we imply that the 4 is uncertain but not the 6. If the result were $L = 3.264 \pm 0.012$ m, we would write the shortened value as $L = 3.26$ m, not as $L = 3.264$ m, because now the uncertainty is in the second decimal instead of the third. In this shorthand method we give only the *significant figures* of the result.

Notice that we always give more information by explicitly writing the error associated with a result than by merely using the significant figures. We can only *estimate* the error in a result from a statement of the significant figures. For example, each of the results with errors listed below has the same shorthand value in terms of the significant figures:

$$\left.\begin{array}{l} 2.682 \ \pm 0.002 \text{ m} \\ 2.682 \ \pm 0.004 \text{ m} \\ 2.6821 \pm 0.0016 \text{ m} \\ 2.6821 \pm 0.0009 \text{ m} \end{array}\right\} \ 2.682 \text{ m}$$

Notice in the last case that, although the error is in the fourth decimal place, ± 0.0009 m, we *round-off* the result to 2.682 m because the error is closer to ± 0.001 m than to ± 0.0001 m. In general, we round-off numbers *upwards* if the last digit is greater than 5, and we round-off downwards if the last digit is less than 5. If the last digit is equal to 5, by convention we round-off to an *even* digit. For example,

$$7.687 \pm 0.012 \text{ m} \rightarrow 7.69 \text{ m}$$

$$3.132 \pm 0.026 \text{ m} \rightarrow 3.13 \text{ m}$$

$$8.274 \pm 0.010 \text{ m} \rightarrow 8.27 \text{ m}$$

$$4.865 \pm 0.018 \text{ m} \rightarrow 4.86 \text{ m}$$

$$2.335 \pm 0.009 \text{ m} \rightarrow 2.34 \text{ m}$$

The placement of the decimal point in a number representing a physical quantity is determined by the *units* that we use for the quantity. For example,

$$17 \text{ mm} = 1.7 \text{ cm} = 0.017 \text{ m} = 0.000 \ 017 \text{ km}$$

Each of these lengths is given to *two* significant figures. That is, we count significant figures from left to right, ignoring all preceding zeros.

$$2.83 \text{ g} \quad \rightarrow 3 \text{ significant figures}$$

$$0.37 \text{ kg} \quad \rightarrow 2 \text{ significant figures}$$

$$0.073 \text{ m} \quad \rightarrow 2 \text{ significant figures}$$

Because we can always express a result in many different units, we must be careful to write the result in a form that exhibits the significant figures in an unambiguous way. For example, if we have a result of 2.17 km, implying that the 7 is the last significant figure, we should not write this as 217,000 cm, which implies that there are *six* significant figures. Instead we should use the powers-of-10 notation and give the result as 2.17×10^5 cm. This procedure allows us automatically to terminate the number of digits with the last significant figure. If the first zero of 217,000 cm were significant, we would write 2.170×10^5 cm.

$$3.72 \times 10^4 \text{ g} \quad \rightarrow 3 \text{ significant figures}$$

$$6.80 \times 10^5 \text{ m} \quad \rightarrow 3 \text{ significant figures}$$

$$0.79 \times 10^{-3} \text{ cm} \rightarrow 2 \text{ significant figures}$$

$$2.300 \times 10^4 \text{ s} \quad \rightarrow 4 \text{ significant figures}$$

EXERCISES	*Express the following results in terms of significant figures:*

1. 4.638 ± 0.002 kg   (Ans. 8)
2. 1.4 ± 0.6 m/s^2 (Ans. 209)
3. $(6.319 \pm 0.004) \times 10^6$ cm   (Ans. 309)
4. $(6.319 \pm 0.008) \times 10^6$ cm   (Ans. 343)
5. 9.52 ± 0.01 m (express in km)   (Ans. 174)
6. 5.356 ± 0.005 m (express in cm)   (Ans. 41)

7. 1.4 ± 0.2 m/s^2 (Ans. 69)
8. If $F = 10.983 \pm 0.003$ N, F probably lies between what two values?   (Ans. 32)
9. How many significant figures are there in the result $F = 10.983$ N?   (Ans. 275)

Round-off each of the following to three significant figures:

10. 5.918 N   (Ans. 266)
11. 3.798 cm   (Ans. 31)
12. 0.05117 m/s   (Ans. 24)

13. 2365.987 N   (Ans. 5)
14. 0.123 cm   (Ans. 202)
15. 6.895×10^6 mi/h   (Ans. 6)

7.4   MULTIPLICATION AND DIVISION OF EXPERIMENTAL NUMBERS

Many physical quantities are determined indirectly by measurements of other quantities. For example, the surface area of a rectangular piece of metal is determined by measuring the length and the width. The velocity of an object is determined by measuring the distance moved and the time interval required for the movement. In each case we have two experimental quantities that must be combined to produce the desired result. Each of the input quantities has associated with it an experimental uncertainty; how do we determine the uncertainty in the result?

Suppose, for example, that we wish to determine the area of a small rectangular sample of metal. Using calipers, we measure the length to be 9.8 mm and the width to be 3.2 mm. To each of these measurements we assign an uncertainty of ± 0.1 mm. That is, our input quantities are

$$l = 9.8 \pm 0.1 \text{ mm (relative error} = 1\%)$$

$$w = 3.2 \pm 0.1 \text{ mm (relative error} = 3\%)$$

with relative errors of $0.1/9.8 = 0.01$ (1%) for l and $0.1/3.2 = 0.03$ (3%) for w. The area is

$$A = l \times w = (9.8 \text{ mm}) \times (3.2 \text{ mm}) = 31.36 \text{ mm}^2$$

What uncertainty do we assign to A? If the last digit of the result (i.e., 6) were significant, then the area would have an error of approximately ± 0.01, or a relative error of approximately $0.01/31.36 = 0.0003$ (0.03%). Surely the area cannot be uncertain by only 0.03% when the length and width have errors of 1% and 3%, respectively! Therefore, not all the figures in the product $l \times w$ are significant. We can see this as follows:

First, we compute the area by taking values for l and w corresponding to the *lower* end of the range of probable error; call this area A^-:

$$A^- = (9.8 - 0.1 \text{ m}) \times (3.2 - 0.1 \text{ mm})$$
$$= (9.7 \text{ mm}) \times (3.1 \text{ mm}) = 30.07 \text{ mm}^2$$

Next, we compute A^+ by taking values for l and w corresponding to the *upper* end of the range of probable error:

$$A^+ = (9.8 + 0.1 \text{ mm}) \times (3.2 + 0.1 \text{ mm})$$
$$= (9.9 \text{ mm}) \times (3.3 \text{ mm}) = 32.67 \text{ mm}^2$$

The extreme spread in these results is

$$A^+ - A^- = 32.67 \text{ mm}^2 - 30.07 \text{ mm}^2 = 2.60 \text{ mm}^2$$

The value of A is 31.36 mm^2, as we found above, and we take for the range of uncertainty in A this extreme spread from A^- to A^+. But A lies halfway between A^- and A^+, so we can state the results as $A \pm \frac{1}{2}(A^+ - A^-)$; that is,

$$A = 31.36 \pm 1.30 \text{ mm}^2$$

or, rounding off the last decimal and calculating the relative error, we find

$$A = 31.4 \pm 1.3 \text{ mm}^2 \text{ (relative error} = 4\%)$$

We can draw the following conclusions from this example. First, the relative error in the result (4%) is *larger* than the relative error in either of the input quantities (1% and 3%). This is easy to understand—we cannot *improve* the precision by simply multiplying two quantities. Second, we notice that the relative error in the area is just the *sum* of the relative errors in the length and the width. This is, in fact, a general rule*:

> Relative error in a product or a quotient equals the sum of the relative errors in the input quantities.
>
> (7.6)

Notice that the rule applies to *quotients* as well as to *products*. (Division, after all, is just multiplication by the reciprocal.)

---

*As stated, this rule does not agree exactly with that derived from a detailed analysis of error processes, but it is sufficiently close to be adequate for all our purposes.

▶ **EXAMPLE 7.4.1**   What is the area (and the uncertainty) of a circle whose radius is 5.2 ± 0.2 cm?
Although we have only one measured quantity in this case, we still have a *product*
to consider:

$$A = \pi r^2 = \pi \times r \times r = \pi \times (5.2 \pm 0.2 \text{ cm}) \times (5.2 \pm 0.2 \text{ cm})$$

The relative error in the radius is 0.2/5.2 = 0.04 (4%); therefore, the relative error
in the area is 2 × 4% = 8%. (There is no error in the quantity π and this multi-
plicative factor does not influence the relative error in the result.) Thus,

$$A = \pi \times (5.2 \text{ cm})^2 \pm 8\% = 8.5 \pm 0.7 \text{ cm}^2$$

In this case we have rounded off the result and retained only one significant figure
in the uncertainty. (Compare with the example above where we carried two significant figures and wrote ±1.3 mm².) Usually, it is sufficient to round off all quantities
to leave one significant figure in the uncertainty of the result.

▶ **EXAMPLE 7.4.2**   In order to determine the speed of a moving object, we measure the time required
for the object to pass between two points which are 1.015 ± 0.002 m apart. The
measured time interval is (3.5 ± 0.1) × 10^{-3} s. What is the speed?
The input quantities are

$$l = 1.015 \pm 0.002 \text{ m (relative error} = 0.2\%)$$

$$t = (3.5 \pm 0.1) \times 10^{-3} \text{ s (relative error} = 3\%)$$

Therefore, calculating l/t and adding the relative errors, we find

$$v = \frac{l}{t} = \frac{1.015 \text{ m}}{3.5 \times 10^{-3} \text{ s}} = (290 \text{ m/s}) \pm 3.2\%$$

so that

$$v = 290 \pm 9 \text{ m/s}$$

In this case, one of the input quantities (the length) is known with considerably
higher precision than the other quantity (the time). Therefore, the uncertainty in the
result is due almost entirely to the timing error. (In fact, we could have ignored the
length error without changing the final error since we retain only one significant
figure in the final error.)

EXERCISES   *What are the uncertainties in each of the areas determined from the following length and width measurements?*

1. $w = 50.1 \pm 0.3$ m, $l = 7.8 \pm 0.3$ m   (Ans. 336)
2. $w = (50 \pm 1) \times 10^{-8}$ cm, $l = (100 \pm 2) \times 10^{-8}$ cm
      (Ans. 77)
3. $w = (2.4 \pm 0.2) \times 10^{7}$ km, $l = (5.2 \pm 0.2) \times 10^{8}$ km
      (Ans. 144)
4. What is the uncertainty in the volume of a sphere that has a radius of 1.01 ± 0.01 m?   (Ans. 124)

5. What is the uncertainty in the area of a square with sides of length 2.3 ± 0.3 cm?   (Ans. 177)
6. An object has a mass of 1.6 ± 0.1 kg and is moving with a velocity of 20.3 ± 0.3 m/s. What is the momentum of the object?   (Ans. 243)
7. Calculate the kinetic energy of the object in Exercise 6 above.   (Ans. 135)

7.5   ADDITION AND SUBTRACTION OF EXPERIMENTAL NUMBERS

Suppose that we prepare a certain mixture by adding 10.2 ± 0.3 g of substance A to 304 ± 3 g of substance B. What will be the mass of the mixture and what will be the uncertainty in the result? The input quantities are

$$m_A = 10.2 \pm 0.3 \text{ g}$$

$$m_B = 304 \pm 3 \text{ g}$$

The relative error in m_A is 3%, whereas the relative error in m_B is 1%. If we were to follow the prescription for computing the uncertainty as discussed in the preceding section, we would conclude that the relative error in $M = m_A + m_B$ is 4%, giving an absolute error of

$$\text{absolute error} = 0.04 \times (10.2 \text{ g} + 304 \text{ g})$$

$$= 0.04 \times 314.2 \text{ g} = 12.6 \text{ g}$$

But this error is considerably larger than the absolute error in either m_A or m_B and, in fact, is even larger than m_A itself! This method of computing the uncertainty is surely wrong.

From the standpoint of the physical process involved in this problem, we can argue as follows. Substance B alone has a mass uncertainty of ± 3 g; adding any amount of substance A will not alter that value, and the added amount of substance A carries its own mass uncertainty to the mixture. We must conclude, therefore, that the absolute error in the mixture is equal to the *sum* of the *absolute errors* in the constituents, or ± 3.3 g in this case. The general rule is*

> *Absolute* error in an *addition* or *subtraction* equals the *sum* of the *absolute* errors in the input quantities.
>
>   **(7.7)**

---

*Again, this rule is not strictly correct but it is entirely sufficient for our purposes.

▶ **EXAMPLE 7.5.1**   The distance from a point P to a point P' is determined in two steps. The individual results are $l_1 = 37.2 \pm 0.2$ m and $l_2 = 43.6 \pm 0.3$ m. What is the distance PP'? Following the general rule, we have

$$
\begin{array}{r}
l_1 = 37.2 \pm 0.2 \text{ m} \\
l_2 = 43.6 \pm 0.3 \text{ m} \\
\hline
PP' = 80.8 \pm 0.5 \text{ m}
\end{array}
$$

▶ **EXAMPLE 7.5.2**   Measurements are made of two masses with the results $m_1 = 48.1 \pm 0.3$ g and $m_2 = 47.6 \pm 0.2$ g. These masses are placed on a beam balance and compared. What is the mass difference between m_1 and m_2? According to the general rule we have

$$
\begin{array}{r}
m_1 = 48.1 \pm 0.3 \text{ g} \\
m_2 = 47.6 \pm 0.2 \text{ g} \\
\hline
\text{difference} = 0.5 \pm 0.5 \text{ g}
\end{array}
$$

That is, the uncertainty in the mass difference is equal to the mass difference itself. This is quite reasonable when we realize that the ranges of uncertainty for the two measurements just meet, and therefore, within the limits of precision of the two measurements, the masses could, in fact, be equal.

EXERCISES   *Compute the following sums and differences and their uncertainties:*

1. $(5.98 \pm 0.01$ km$) + (6.32 \pm 0.02$ km$)$ (Ans. 286)
2. $(5.98 \pm 0.01$ km$) - (6.32 \pm 0.02$ km$)$ (Ans. 145)
3. $(16.323 \pm 0.003$ g$) + (12.401 \pm 0.001$ g$)$
    (Ans. 240)
4. $(2.5 \pm 0.1$ m$) - (2.0 \pm 0.1$ m$)$ (Ans. 315)
5. What is the perimeter of a rectangle with two sides

of length $l_1 = 10.3 \pm 0.1$ m, and two sides of length $l_2 = 20.2 \pm 0.2$ m?   (Ans. 150)
6. Two masses ($m_1 = 10.1 \pm 0.1$ g and $m_2 = 323.2 \pm 0.2$ g) are placed in one pan of a balance beam. Calculate the mass of the weight that must be placed in the other pan in order to balance the beam.   (Ans. 70)

7.6   COUNTING STATISTICS

Suppose that we use a Geiger counter to record the radioactive emissions from a sample of uranium ore. We count the number of detected particles for some predetermined period of time—for example, 1 minute. If we repeat the measurement many times, always keeping conditions the same, we find that we do not record exactly the same number of counts in each 1-min interval. We might obtain a series of counts such as

$$1013, 984, 1021, 1003, 995, 978, 1019, 989$$

The reason for the variation in the number of counts per minute is that there is no way to predict the instant when any given radioactive nucleus will decay. We can only specify the *probability* that a given nucleus will decay during a certain specified time interval. When we deal with large numbers of radioactive nuclei (as in a sample of uranium ore), the decay particles are emitted from the sample spaced at *random* intervals of time. Therefore, we cannot expect the same number of decay events to take place within two identical but short intervals of time. If we lengthen the time interval for counting, we again find that the number of counts varies from one measurement to the next; however, the *relative* (or percentage) variation is *smaller* than for the shorter time interval.

The general rule for the counting of radioactive decay events (or for *any* type of process that occurs with random time spacing) is the following. If a large number of measurements is made and the *average* number of counts per minute is found to be N, then the individual measurements will be distributed in such a way that approximately one-half of the individual results will lie in the range $N \pm \sqrt{N}$. That is, each individual measurement has associated with it a kind of "probable error" equal to $\sqrt{N}$.

We can immediately see that longer counting times provide increased precision in measuring the rate of occurrence of decay events. If we wish to know the decay rate to $\pm 10\%$, then we need count only long enough to record 100 counts, for then we have $100 \pm \sqrt{100} = 100 \pm 10 = 100 \pm 10\%$. However, if we wish to increase the precision by a factor of 10 to $\pm 1\%$, we must count 100 times as long in order to obtain $10{,}000 \pm \sqrt{10{,}000} = 10{,}000 \pm 100 = 10{,}000 \pm 1\%$.

▶ **EXAMPLE 7.6.1**   A detector records 41,376 counts from a certain radioactive sample in a period of 7.2 min. What is the decay rate of the material?

The uncertainty in the number of decays is $\sqrt{N}$, where $N = 41{,}376$. Therefore,

$$N \pm \sqrt{N} = 41{,}376 \pm \sqrt{41{,}376} = 41{,}376 \pm 203$$

and the decay rate is

$$\text{decay rate} = \frac{41{,}376 \pm 203}{7.2 \times 60 \text{ s}} = 95.8 \pm 0.5 \text{ s}^{-1}$$

The precision of the result is

$$\text{precision} = \frac{\sqrt{N}}{N} = \frac{1}{\sqrt{N}} = \frac{1}{203} = 0.5\%$$

EXERCISES

1. In a one-hour interval, a detector records 635,237 counts from a certain radioactive sample. What is the decay rate of the sample?   (Ans. 328)
2. What is the precision of the result in Exercise 1 above?   (Ans. 2)
3. A birdwatcher records 230 birds in a certain meadow between 10 AM and 11 AM. The next day, during the same hour, only 110 birds were recorded. What can you conclude from this observation?   (Ans. 167)
4. Is there any difference (from the standpoint of the uncertainty in the final result) of taking ten 1-min counting periods or one 10-min counting period to determine an average counting rate?   (Ans. 340)

A

Some Useful Tables

TABLE I   Values of the Trigonometric Functions

deg	rad	sin θ	cos θ	tan θ	deg	rad	sin θ	cos θ	tan θ
0	0.000	0.000	1.000	0.000					
1	0.017	0.017	1.000	0.017	46	0.803	0.719	0.695	1.036
2	0.035	0.035	0.999	0.035	47	0.820	0.731	0.682	1.072
3	0.052	0.052	0.999	0.052	48	0.838	0.743	0.669	1.111
4	0.070	0.070	0.998	0.070	49	0.855	0.755	0.656	1.150
5	0.087	0.087	0.996	0.087	50	0.873	0.766	0.643	1.192
6	0.105	0.105	0.995	0.105	51	0.890	0.777	0.629	1.235
7	0.122	0.122	0.993	0.123	52	0.908	0.788	0.616	1.280
8	0.140	0.139	0.990	0.141	53	0.925	0.799	0.602	1.327
9	0.157	0.156	0.988	0.158	54	0.942	0.809	0.588	1.376
10	0.175	0.174	0.985	0.176	55	0.960	0.819	0.574	1.428
11	0.192	0.191	0.982	0.194	56	0.977	0.829	0.559	1.483
12	0.209	0.208	0.978	0.213	57	0.995	0.839	0.545	1.540
13	0.227	0.225	0.974	0.231	58	1.012	0.848	0.530	1.600
14	0.244	0.242	0.970	0.249	59	1.030	0.857	0.515	1.664
15	0.262	0.259	0.966	0.268	60	1.047	0.866	0.500	1.732
16	0.279	0.276	0.961	0.287	61	1.065	0.875	0.485	1.804
17	0.297	0.292	0.956	0.306	62	1.082	0.883	0.470	1.881
18	0.314	0.309	0.951	0.325	63	1.100	0.891	0.454	1.963
19	0.332	0.326	0.946	0.344	64	1.117	0.899	0.438	2.050
20	0.349	0.342	0.940	0.364	65	1.134	0.906	0.423	2.145
21	0.367	0.358	0.934	0.384	66	1.152	0.914	0.407	2.246
22	0.384	0.375	0.927	0.404	67	1.169	0.921	0.391	2.356
23	0.401	0.391	0.921	0.424	68	1.187	0.927	0.375	2.475
24	0.419	0.407	0.914	0.445	69	1.204	0.934	0.358	2.605
25	0.436	0.423	0.906	0.466	70	1.222	0.940	0.342	2.747
26	0.454	0.438	0.899	0.488	71	1.239	0.946	0.326	2.904
27	0.471	0.454	0.891	0.510	72	1.257	0.951	0.309	3.078
28	0.489	0.470	0.883	0.532	73	1.274	0.956	0.292	3.271
29	0.506	0.485	0.875	0.554	74	1.292	0.961	0.276	3.487
30	0.524	0.500	0.866	0.577	75	1.309	0.966	0.259	3.732
31	0.541	0.515	0.857	0.601	76	1.326	0.970	0.242	4.011
32	0.559	0.530	0.848	0.625	77	1.344	0.974	0.225	4.331
33	0.576	0.545	0.839	0.649	78	1.361	0.978	0.208	4.705
34	0.593	0.559	0.829	0.675	79	1.379	0.982	0.191	5.145
35	0.611	0.574	0.819	0.700	80	1.396	0.985	0.174	5.671
36	0.628	0.588	0.809	0.727	81	1.414	0.988	0.156	6.314
37	0.646	0.602	0.799	0.754	82	1.431	0.990	0.139	7.115
38	0.663	0.616	0.788	0.781	83	1.449	0.993	0.122	8.144
39	0.681	0.629	0.777	0.810	84	1.466	0.995	0.105	9.514
40	0.698	0.643	0.766	0.839	85	1.484	0.996	0.087	11.43
41	0.716	0.656	0.755	0.869	86	1.501	0.998	0.070	14.30
42	0.733	0.669	0.743	0.900	87	1.518	0.999	0.052	19.08
43	0.750	0.682	0.731	0.933	88	1.536	0.999	0.035	28.64
44	0.768	0.695	0.719	0.966	89	1.553	1.000	0.017	57.29
45	0.785	0.707	0.707	1.000	90	1.571	1.000	0.000	∞

TABLE II   Squares, Cubes, and Roots*

n'	n^2	$\sqrt{n}$	$\sqrt{10n}$	n^3	$\sqrt[3]{n}$	$\sqrt[3]{10n}$	$\sqrt[3]{100n}$
1	1	1.000 000	3.162 278	1	1.000 000	2.154 435	4.641 589
2	4	1.414 214	4.472 136	8	1.259 921	2.714 418	5.848 035
3	9	1.732 051	5.477 226	27	1.442 250	3.107 233	6.694 330
4	16	2.000 000	6.324 555	64	1.587 401	3.419 952	7.368 063
5	25	2.236 068	7.071 068	125	1.709 976	3.684 031	7.937 005
6	36	2.449 490	7.745 967	216	1.817 121	3.914 868	8.434 327
7	49	2.645 751	8.366 600	343	1.912 931	4.121 285	8.879 040
8	64	2.828 427	8.944 272	512	2.000 000	4.308 869	9.283 178
9	81	3.000 000	9.486 833	729	2.080 084	4.481 405	9.654 894
10	100	3.162 278	10.00000	1 000	2.154 435	4.641 589	10.00000
11	121	3.316 625	10.48809	1 331	2.223 980	4.791 420	10.32280
12	144	3.464 102	10.95445	1 728	2.289 428	4.932 424	10.62659
13	169	3.605 551	11.40175	2 197	2.351 335	5.065 797	10.91393
14	196	3.741 657	11.83216	2 744	2.410 142	5.192 494	11.18689
15	225	3.872 983	12.24745	3 375	2.466 212	5.313 293	11.44714
16	256	4.000 000	12.64911	4 096	2.519 842	5.428 835	11.69607
17	289	4.123 106	13.03840	4 913	2.571 282	5.539 658	11.93483
18	324	4.242 641	13.41641	5 832	2.620 741	5.646 216	12.16440
19	361	4.358 899	13.78405	6 859	2.668 402	5.748 897	12.38562
20	400	4.472 136	14.14214	8 000	2.714 418	5.848 035	12.59921
21	441	4.582 576	14.49138	9 261	2.758 924	5.943 922	12.80579
22	484	4.690 416	14.83240	10 648	2.802 039	6.036 811	13.00591
23	529	4.795 832	15.16575	12 167	2.843 867	6.126 926	13.20006
24	576	4.898 979	15.49193	13 824	2.884 499	6.214 465	13.38866
25	625	5.000 000	15.81139	15 625	2.924 018	6.299 605	13.57209
26	676	5.099 020	16.12452	17 576	2.962 496	6.382 504	13.75069
27	729	5.196 152	16.43168	19 683	3.000 000	6.463 304	13.92477
28	784	5.291 503	16.73320	21 952	3.036 589	6.542 133	14.09460
29	841	5.385 165	17.02939	24 389	3.072 317	6.619 106	14.26043
30	900	5.477 226	17.32051	27 000	3.107 233	6.694 330	14.42250
31	961	5.567 764	17.60682	29 791	3.141 381	6.767 899	14.58100
32	1 024	5.656 854	17.88854	32 768	3.174 802	6.839 904	14.73613
33	1 089	5.744 563	18.16590	35 937	3.207 534	6.910 423	14.88806
34	1 156	5.830 952	18.43909	39 304	3.239 612	6.979 532	15.03695
35	1 225	5.916 080	18.70829	42 875	3.271 066	7.047 299	15.18294
36	1 296	6.000 000	18.97367	46 656	3.301 927	7.113 787	15.32619
37	1 369	6.082 763	19.23538	50 653	3.332 222	7.179 054	15.46680
38	1 444	6.164 414	19.49359	54 872	3.361 975	7.243 156	15.60491
39	1 521	6.244 998	19.74842	59 319	3.391 211	7.306 144	15.74061
40	1 600	6.324 555	20.00000	64 000	3.419 952	7.368 063	15.87401
41	1 681	6.403 124	20.24846	68 921	3.448 217	7.428 959	16.00521
42	1 764	6.480 741	20.49390	74 088	3.476 027	7.488 872	16.13429
43	1 849	6.557 439	20.73644	79 507	3.503 398	7.547 842	16.26133
44	1 936	6.633 250	20.97618	85 184	3.530 348	7.605 905	16.38643
45	2 025	6.708 204	21.21320	91 125	3.556 893	7.663 094	16.50964
46	2 116	6.782 330	21.44761	97 336	3.583 048	7.719 443	16.63103
47	2 209	6.855 655	21.67948	103 823	3.608 826	7.774 980	16.75069
48	2 304	6.928 203	21.90890	110 592	3.634 241	7.829 735	16.86865
49	2 401	7.000 000	22.13594	117 649	3.659 306	7.883 735	16.98499

*See footnote on page 223.

TABLE II   Squares, Cubes, and Roots*—Continued

n'	n^2	$\sqrt{n}$	$\sqrt{10n}$	n^3	$\sqrt[3]{n}$	$\sqrt[3]{10n}$	$\sqrt[3]{100n}$
50	2 500	7.071 068	22.36068	125 000	3.684 031	7.937 005	17.09976
51	2 601	7.141 428	22.58318	132 651	3.708 430	7.989 570	17.21301
52	2 704	7.211 103	22.80351	140 608	3.732 511	8.041 452	17.32478
53	2 809	7.280 110	23.02173	148 877	3.756 286	8.092 672	17.43513
54	2 916	7.348 469	23.23790	157 464	3.779 763	8.143 253	17.54411
55	3 025	7.416 198	23.45208	166 375	3.802 952	8.193 213	17.65174
56	3 136	7.483 315	23.66432	175 616	3.825 862	8.242 571	17.75808
57	3 249	7.549 834	23.87467	185 193	3.848 501	8.291 344	17.86316
58	3 364	7.615 773	24.08319	195 112	3.870 877	8.339 551	17.96702
59	3 481	7.681 146	24.28992	205 379	3.892 996	8.387 207	18.06969
60	3 600	7.745 967	24.49490	216 000	3.914 868	8.434 327	18.17121
61	3 721	7.810 250	24.69818	226 981	3.936 497	8.480 926	18.27160
62	3 844	7.874 008	24.89980	238 328	3.957 892	8.527 019	18.37091
63	3 969	7.937 254	25.09980	250 047	3.979 057	8.572 619	18.46915
64	4 096	8.000 000	25.29822	262 144	4.000 000	8.617 739	18.56636
65	4 225	8.062 258	25.49510	274 625	4.020 726	8.662 391	18.66256
66	4 356	8.124 038	25.69047	287 496	4.041 240	8.706 588	18.75777
67	4 489	8.185 353	25.88436	300 763	4.061 548	8.750 340	18.85204
68	4 624	8.246 211	26.07681	314 432	4.081 655	8.793 659	18.94536
69	4 761	8.306 624	26.26785	328 509	4.101 566	8.836 556	19.03778
70	4 900	8.366 600	26.45751	343 000	4.121 285	8.879 040	19.12931
71	5 041	8.426 150	26.64583	357 911	4.140 818	8.921 121	19.21997
72	5 184	8.485 281	26.83282	373 248	4.160 168	8.962 809	19.30979
73	5 329	8.544 004	27.01851	389 017	4.179 339	9.004 113	19.39877
74	5 476	8.602 325	27.20294	405 224	4.198 336	9.045 042	19.48695
75	5 625	8.660 254	27.38613	421 875	4.217 163	9.085 603	19.57434
76	5 776	8.717 798	27.56810	438 976	4.235 824	9.125 805	19.66095
77	5 929	8.774 964	27.74887	456 533	4.254 321	9.165 656	19.74681
78	6 084	8.831 761	27.92848	474 552	4.272 659	9.205 164	19.83192
79	6 241	8.888 194	28.10694	493 039	4.290 840	9.244 335	19.91632
80	6 400	8.944 272	28.28427	512 000	4.308 869	9.283 178	20.00000
81	6 561	9.000 000	28.46050	531 441	4.326 749	9.321 698	20.08299
82	6 724	9.055 385	28.63564	551 368	4.344 481	9.359 902	20.16530
83	6 889	9.110 434	28.80972	571 787	4.362 071	9.397 796	20.24694
84	7 056	9.165 151	28.98275	592 704	4.379 519	9.435 388	20.32793
85	7 225	9.219 544	29.15476	614 125	4.396 830	9.472 682	20.40828
86	7 396	9.273 618	29.32576	636 056	4.414 005	9.509 685	20.48800
87	7 569	9.327 379	29.49576	658 503	4.431 048	9.546 403	20.56710
88	7 744	9.380 832	29.66479	681 472	4.447 960	9.582 840	20.64560
89	7 921	9.433 981	29.83287	704 969	4.464 745	9.619 002	20.72351
90	8 100	9.486 833	30.00000	729 000	4.481 405	9.654 894	20.80084
91	8 281	9.539 392	30.16621	753 571	4.497 941	9.690 521	20.87759
92	8 464	9.591 663	30.33150	778 688	4.514 357	9.725 888	20.95379
93	8 649	9.643 651	30.49590	804 357	4.530 655	9.761 000	21.02944
94	8 836	9.695 360	30.65942	830 584	4.546 836	9.795 861	21.10454

*See footnote on page 223.

TABLE II   Squares, Cubes, and Roots*—Continued

n'	n^2	$\sqrt{n}$	$\sqrt{10n}$	n^3	$\sqrt[3]{n}$	$\sqrt[3]{10n}$	$\sqrt[3]{100n}$
95	9 025	9.746 794	30.82207	857 375	4.562 903	9.830 476	21.17912
96	9 216	9.797 959	30.98387	884 736	4.578 857	9.864 848	21.25317
97	9 409	9.848 858	31.14482	912 673	4.594 701	9.898 983	21.32671
98	9 604	9.899 495	31.30495	941 192	4.610 436	9.932 884	21.39975
99	9 801	9.949 874	31.46427	970 299	4.626 065	9.966 555	21.47229
100	10 000	10.00000	31.62278	1 000 000	4.641 589	10.00000	21.54435

*Roots of numbers other than those given in this table may be determined from the following relations:

Square Roots:

$$\sqrt{1000n} = 10\sqrt{10n}; \; \sqrt{100n} = 10\sqrt{n}; \; \sqrt{\frac{n}{10}} = \frac{\sqrt{10n}}{10}; \; \sqrt{\frac{n}{100}} = \frac{\sqrt{n}}{10}; \; \sqrt{\frac{n}{1000}} = \frac{\sqrt{10n}}{100}$$

Cube Roots:

$$\sqrt[3]{100,000n} = 10\sqrt[3]{100n}; \; \sqrt[3]{10,000n} = 10\sqrt[3]{10n}; \; \sqrt[3]{1000n} = 10\sqrt[3]{n}; \; \sqrt[3]{\frac{n}{100}} =$$

$$\frac{\sqrt[3]{10n}}{10}; \; \sqrt[3]{\frac{n}{1000}} = \frac{\sqrt[3]{n}}{10}$$

TABLE III   Exponential Functions

x	e^x	e^{-x}	x	e^x	e^{-x}
0	1.0000	1.0000	1.6	4.9530	0.2019
0.1	1.1052	0.9048	1.7	5.4739	0.1827
0.2	1.2214	0.8187	1.8	6.0496	0.1653
0.3	1.3499	0.7408	1.9	6.6859	0.1496
0.4	1.4918	0.6703	2.0	7.3891	0.1353
0.5	1.6487	0.6065	2.5	12.182	0.08208
0.6	1.8221	0.5488	3.0	20.086	0.04979
0.7	2.0138	0.4966	3.5	33.115	0.03020
0.8	2.2255	0.4493	4.0	54.598	0.01832
0.9	2.4596	0.4066	4.5	90.017	0.01111
1.0	2.7183	0.3679	5.0	148.41	0.006738
1.1	3.0042	0.3329	6.0	403.43	0.002479
1.2	3.3201	0.3012	7.0	1096.6	0.0009119
1.3	3.6693	0.2725	8.0	2981.0	0.0003355
1.4	4.0552	0.2466	9.0	8103.1	0.0001234
1.5	4.4817	0.2231	10.0	22026.	0.0000454

TABLE IV   Logarithms to the Base 10

The common logarithms[1] of numbers other than those given in the table below may be determined from the following relations.[2]

$$\log 10n = \log n + 1; \log 100n = \log n + 2; \log 1000n = \log n + 3$$

$$\log \frac{n}{10} = \log n - 1; \log \frac{n}{100} = \log n - 2; \log \frac{n}{1000} = \log n - 3.$$

n	$\log n$	n	$\log n$	n	$\log n$
1.0	0.0000	4.0	0.6021	7.0	0.8451
1.1	0.0414	4.1	0.6128	7.1	0.8513
1.2	0.0792	4.2	0.6232	7.2	0.8573
1.3	0.1139	4.3	0.6335	7.3	0.8633
1.4	0.1461	4.4	0.6435	7.4	0.8692
1.5	0.1761	4.5	0.6532	7.5	0.8751
1.6	0.2041	4.6	0.6628	7.6	0.8808
1.7	0.2304	4.7	0.6721	7.7	0.8865
1.8	0.2553	4.8	0.6812	7.8	0.8921
1.9	0.2788	4.9	0.6902	7.9	0.8976
2.0	0.3010	5.0	0.6990	8.0	0.9031
2.1	0.3222	5.1	0.7076	8.1	0.9085
2.2	0.3424	5.2	0.7160	8.2	0.9138
2.3	0.3617	5.3	0.7243	8.3	0.9191
2.4	0.3802	5.4	0.7324	8.4	0.9243
2.5	0.3979	5.5	0.7404	8.5	0.9294
2.6	0.4150	5.6	0.7482	8.6	0.9345
2.7	0.4314	5.7	0.7559	8.7	0.9395
2.8	0.4472	5.8	0.7634	8.8	0.9445
2.9	0.4624	5.9	0.7709	8.9	0.9494
3.0	0.4771	6.0	0.7782	9.0	0.9542
3.1	0.4914	6.1	0.7853	9.1	0.9590
3.2	0.5051	6.2	0.7924	9.2	0.9638
3.3	0.5185	6.3	0.7993	9.3	0.9685
3.4	0.5315	6.4	0.8062	9.4	0.9731
3.5	0.5441	6.5	0.8129	9.5	0.9777
3.6	0.5563	6.6	0.8195	9.6	0.9823
3.7	0.5682	6.7	0.8261	9.7	0.9868
3.8	0.5798	6.8	0.8325	9.8	0.9912
3.9	0.5911	6.9	0.8388	9.9	0.9956

[1]Logarithms to the base 10.

[2]These relations follow from the identities $\log AB = \log A + \log B$, and $\log \frac{A}{B} = \log A - \log B$.

TABLE V   Natural Logarithms

x	$\ln x = \log_e x$	x	$\ln x = \log_e x$
0	$-\infty$	1.9	0.6419
0.01	-4.6052	2.0	0.6932
0.02	-3.9120	2.5	0.9163
0.03	-3.5066	3.5	1.0986
0.04	-3.2189	4.0	1.2528
0.05	-2.9957	4.5	1.5041
0.06	-2.8134	5.0	1.6094
0.07	-2.6593	5.5	1.7048
0.08	-2.5257	6.0	1.7918
0.09	-2.4080	6.5	1.8718
0.1	-2.3026	7.0	1.9459
0.2	-1.6094	7.5	2.0149
0.3	-1.2040	8.0	2.0794
0.4	-0.9163	8.5	2.1401
0.5	-0.6932	9.0	2.1972
0.6	-0.5108	9.5	2.2513
0.7	-0.3567	10.0	2.3026
0.8	-0.2231	11.0	2.3979
0.9	-0.1054	12.0	2.4849
1.0	0	13.0	2.5650
1.1	0.0953	14.0	2.6391
1.2	0.1823	15.0	2.7081
1.3	0.2624	20.0	2.9957
1.4	0.3365	25.0	3.2189
1.5	0.4055	30.0	3.4012
1.6	0.4700	35.0	3.5554
1.7	0.5306	40.0	3.6889
1.8	0.5878	45.0	3.8067
		50.0	3.9120

TABLE VI   Physical Data

Physical Constants

Velocity of light (in vacuum)	$c = 2.998 \times 10^8$ m/s
Charge of electron	$e = 1.602 \times 10^{-19}$ C
Planck's constant	$h = 6.626 \times 10^{-34}$ J·s
Boltzmann's constant	$k = 1.381 \times 10^{-23}$ J/K
Gas constant	$R = 8.314$ J/mol·K
Stefan-Boltzmann constant	$\sigma = 5.670 \times 10^{-8}$ (W/m^2)/K^4
Avogadro's number	$N_0 = 6.022 \times 10^{23}$ mol^{-1}
Gravitational constant	$G = 6.673 \times 10^{-11}$ N·m^2/kg^2
"Standard gravity"	$g = 9.807$ m/s^2
Electric force constant	$K = 8.988 \times 10^9$ N·m^2/C^2
Permeability of free space	$\mu_0 = 4\pi \times 10^{-7}$ N/A^2
Electron mass	$m_e = 9.110 \times 10^{-31}$ kg
Proton mass	$m_p = 1.673 \times 10^{-27}$ kg
Neutron mass	$m_n = 1.675 \times 10^{-27}$ kg

Astronomical Data

Earth-Sun distance (1 astronomical unit, 1 A.U.)	$= 1.496 \times 10^{11}$ m
Earth-Moon distance	$= 3.844 \times 10^8$ m
Radius of Sun	$= 6.960 \times 10^8$ m
Radius of Earth	$= 6.378 \times 10^6$ m
Radius of Moon	$= 1.738 \times 10^6$ m
Mass of Sun	$= 1.989 \times 10^{30}$ kg
Mass of Earth	$= 5.974 \times 10^{24}$ kg
Mass of Moon	$= 7.35 \times 10^{22}$ kg
Average orbital speed of Earth	$= 2.98 \times 10^4$ m/s
	$\cong 30$ km/s
1 Light Year (L.Y.)	$= 9.461 \times 10^{15}$ m

Conversion Factors

1 in	$= 2.54$ cm (exactly)
1 mi	$= 1609$ m
1 yd	$= 0.9144$ m (exactly)
1 m	$= 39.37$ in. $= 1.094$ yd
1 Å	$= 10^{-10}$ m
1 ft/s	$= 0.3048$ m/s
1 mi/h	$= 0.4470$ m/s
1 lb	$= 0.4536$ kg
1 Cal	$= 4186$ J
1 hp	$= 745.7$ W
1 atm	$= 1.013 \times 10^5$ N/m^2 $= 760$ torr
1 y	$= 3.156 \times 10^7$ s

Answers to Exercises

1. $\frac{1}{2} v_0 t_b$

2. Precision $= \dfrac{1}{\sqrt{N}} = \dfrac{1}{797} = 0.13\%$

3. $a(t) = \dfrac{\alpha}{m} t^{1/3}$

$v(t) = v_0 + \dfrac{3\alpha}{2m} t^{2/3}$

$x(t) = x_0 + v_0 t + \dfrac{9\alpha}{10m} t^{5/3}$

4. No

5. 2.37×10^3 N

6. 6.90×10^6 mi/h

7. 2.2×10^9

8. 4.638 kg

9. $t = 4$ s

10. 69.1°

11. 35.2°

12. $\mathbf{F} = 33\,\boldsymbol{j}$ (in newtons)

13. $D = 12\ m$

14. $W_{ab} = \dfrac{F_0}{\alpha}(e^{\alpha x b} - 1)$

15. $t^3 + t^8 + C$

17. 0.0018

18. $v = r\omega$

19. 0

20. 3

21. 256

22. 10 N

23. 1.414 cm

24. 0.0512 m/s

25. The null vector

26. Down

27. 10,400 m/s^2

28. $\mathbf{v}_R$ is a velocity of 632 km/h directed 18.4° east of due north.

29. 43.58 kg·m^2/s

30. 0.9848

31. 3.80 cm

32. 10.98 N and 10.986 N

33. 1.07×10^4

35. 2.45×10^3 J

36. 1

37. 0.8980

38. $\omega = 6.28$ rad/s

39. A displacement of 450 m directed 45° north of due west.

40. $y = \frac{1}{2}x + 2$

41. 536 cm

42. 7.07

43. 0.9798

44. 53°

45. 9.798 m

46. $D = \frac{a}{6}(1 - \cos 6t)$

47. 229.2

48. $7\frac{1}{3}$

49. All except the second; the last one.

51. 5.58

52. $-\sin^3 t + 2\cos^2 t \sin t$

53. 7.3×10^{-5} rad/s

54. 0.6 ppm

55. -3

56. 1.6×10^{-6} J

57. N·m²/kg²

58. 3.74

59. $x = \frac{16}{3} m; t = \frac{4}{3} s$

60. 65°

61. 0.368

62. 0.9986

63. $-\frac{1}{3}\ln|\cos 3t| - \ln t + C$

64. Square

65. -17.343

66. 14.9 m

67. $\mathbf{a} = t^{-1}\mathbf{v} - 2\alpha^2 t^2 \mathbf{r}$

68. $k = 3.95 \times 10^2$ g/s²

69. 1.4 m/s²

70. $m = 333.2 \pm 0.3$ g

71. 1.4 g/cm³

73. $\sec^2 t$

74. 521.1 g

75. 10^8

76. $-\frac{5}{128}b^4$

77. $\pm 4\% = \pm 200 \times 10^{-16}$ cm²

78. 0.25

79. $v = -70\pi \sin(10\pi t)$ cm/s

80. $W_{ab} = \frac{1}{4}kx_b^4$

81. $x = \frac{a}{c} + \frac{b}{c}$

82. $W_{ab} = mgh$

83. $\cos t \cos 2t - 2 \sin t \sin 2t$

84. 12 **F** is a force of 12000 N directed vertically downward.

85. $-\frac{1}{\alpha}(e^{-\alpha t} + e^{\alpha t}) + C$

86.

Time t (seconds)	Displacement x (km)
0	0
1	3
2	6
3	9
4	12
5	15
6	18
7	21
8	24
9	27
10	30

87. 10,000,000

88. 2.7×10^{34} kg·m²/s

90. $D = \frac{-2t}{t + 1}$

91. 10^{-2}

92. 3.35×10^4 g

93. 2×10^3

94. 4

95. 8°

96. $4t^3 \sin 6t + 6t^4 \cos 6t$

97. 30

98. 92.3521

99. $1 + \frac{1}{2} + \frac{1}{3} + \cdots$

100. -0.9367

101. 6×10^{10}

102. 10^9

103. $x = \frac{8}{3}$

104. 2×10^{-6}

105. $a(t) = \frac{F_0}{m} e^{2t}$

    $v(t) = v_0 + \frac{F_0}{2m}(e^{2t} - 1)$

    $x(t) = x_0 - \left(v_0 - \frac{F_0}{2m}\right)t + \frac{F_0}{4m}(e^{2t} - 1)$

106. $x = v_0 t$, where $v_0 = 0.25$ m/s

107. 7

108. 0.969

109. 6

110. 144

111. -0.1736

112. 0.9325

113. $\mathbf{r} \cdot \mathbf{v} = 0$

115. 10^9

116. The displacement x (in km) increases with time t (in s) at a constant rate of $v_0 = 3$ km/s.

117. $\mathbf{F}$ is a force of 22.4 N directed 153.4° from the positive X-axis.

118. $<$

119. 7.87 s

120. 0.9986

121. 10^5

122. $A = 20$ m; $\theta = 90°$; $A_x = 0$; $A_y = 20$ m.

123. Zero

124. $\pm 3\% = \pm 0.13$ m^3

125. 10.9°

126. $x = 21$

127. $W_{ab} = \frac{F_0}{\alpha}(1 - \cos \alpha x_b)$

128. 0

129. $\frac{2}{7} t^{7/2} + C$

130. 1.57×10^5 g

131. 3.7 ppm

132. 0.0873

133. 24°

134. $-6\,\mathbf{F}$ is a force of 6000 N directed vertically upward.

135. 330 ± 30 J

136. $\mathbf{F}$ is a force of 22.4 N directed 63.4° from the positive X-axis.

137. Straight line

138. $\mathbf{v}(t) = -2\alpha r t\,(\sin \alpha t^2\,\mathbf{i} - \cos \alpha t^2\,\mathbf{j})$

139. 1

140. 5120 m/s^2

141. 15 J

142. -1

143. $>$

144. $\pm 12\% = \pm 1.5 \times 10^{15}$ km^2

145. -0.34 ± 0.03 km

146. 10^{-6}

147. $\mathbf{v} = -23\,\mathbf{j}$ (in m/s)

148. $\frac{68}{67}$

150. 61.0 ± 0.6 m

151. $A = 10$ N; $\theta = 0°$; $A_x = 10$ N; $A_y = 0$.

152. 5.39 m

153. 0.995 m

154. Up

155. 125

156. $0.301n$

157. 0.6 s

158. 166.7 m

160. 4.24 cm

161. 2560 m/s

162. 25.4

163. $A = 10$ cm

164. 331×10^9 dollars

165. $1.5\,\mathbf{v}$ is a velocity of 45 mi/h directed 45° north of due east.

166. 0.5

167. Since 110 lies far outside the range $230 \pm \sqrt{230} = 230 \pm 15$, the weather conditions for the count probably were different.

168. 77°

169. 67°

170. $(x, y, z) = (2.5, 4.33, 8.66)$

171. 0.968

172. $-30\,\mathbf{F}$ is a force of 3×10^4 N directed vertically upward.

173. 5134 J

174. 9.52×10^{-3} km

175. 403.4

176. 100 **F** is a force of 10^5 N directed vertically downward.

177. $\pm 26\% = \pm 1.4$ cm^2

178. A displacement of 270 m directed 45° north of due west.

179. $x = \frac{1}{6}, 7$

180. $-6/t^7$

181. $t = 2$ s

182. 10^5

183. 4.07×10^{16} m

184. -0.5446

185. $D = -15$ m

186. $x = \frac{20}{13}$ m; $y = -\frac{17}{13}$ m

187. $-\cos 6t + C$

188. $5120\, t_0^3$ m/s^2

189. 5 m

190. $D = \frac{5}{a}(1 - e^{-at})$

191. 10.1×10^3 g

192. 66.4°

193. 2.24 m

194. $x = 6, -4$

195. 10^{-8}

196. $x_0 = 5$ cm

197. 0.0524

198. 6.68×10^8 J

199. ~64

200. 4.24 cm^2/s^2

201. -1

202. 0.123 cm

203. 0.258

204. $\mathbf{L} = 2\alpha m r^2 t\, \mathbf{k}$

205. 23.6

206. $T = 0.2$ s

207. -3

208. 0.2%

209. 1 m/s^2

210. 0

211. 87.1 m/s

212. A displacement of 45 m directed 45° south of due east.

213. 0.35%

214. $(r, \theta, z) = (2, 45°, 5)$

215. $\dfrac{1}{x} - \dfrac{1}{x^2}$

216. 0.2588

217. $A = 320$ m/s; $\theta = 96°$; $A_x = -33.4$ m/s; $A_y = 318.2$ m/s.

218. $a(t) = \dfrac{F_0}{m} \sin t$

$v(t) = v_0 + \dfrac{F_0}{m}(1 - \cos t)$

$x(t) = x_0 + \left(v_0 + \dfrac{F_0}{m}\right)t - \dfrac{F_0}{m} \sin t$

220. 0.08%

221. $v = \dfrac{2a}{5m}\, t^{5/2}$

222. $-\dfrac{1}{2} v_0 t_b$

223. 0

224. $a = -40\pi^2 \cos(2\pi t + \pi/3)$ cm/s^2

225. $y = \dfrac{13}{7} x$

226. It is possible to represent an event in a conventional plot only if one of the space coordinates is suppressed (leaving, for example, x, y, and t).

227. ~10^{57}

228. 4×10^8 m

230. $W_{ab} = mgh$

231. $12t^{11}$

232. 60

233. 33.3 m/s

234. 173.2 m^2

235. $x = 2$

236. 1.398

237. 0.301

238. $2560\, t_0^4$ m/s

239. $x = 7 \cos(10\pi t)$ cm

240. 28.724 ± 0.004 g

241. Zero

242. 28

243. 32.5 ± 2.5 kg·m/s

244. 91.44

246. 45°

247. $x = \frac{3}{2} + \frac{1}{6}(b - a)$

249. 40.2

250. 1.67×10^{-27} kg

251. -0.1736

252. 500 kg·m²/s

253. $x = \frac{7}{2}, -1$

255. $\omega = 31.4$ rad/s

256. $v_0 t_b$

257. $A_x = 17.68$ m
$A_y = 17.68$ m
$B_x = 0$
$B_y = 25$ m
$R_x = 17.68$ m
$R_y = 42.68$ m
$R = 46.2$ m
$\theta_R = 67.5°$

258. 3.6×10^3 m

259. $12/t^8$

260. $\frac{15}{16}$

261. $v_0 = -54.4$ cm/s

262. 3

263. Up

264. $-\sin t$

265. $t^3 - t \ln 6t + t + C$

266. 5.92 N

267. $1 - x$

268. $t = 2$s

269. 3.2 ppm

270. 1.13×10^7 g

271. -2700 m

272. Down

273. $\phi = \pi/3 = 60°$

274. $3.156 \times 10^8 \cong \pi \times 10^8$

275. Five

276. $1 + 3x + 3x^2 + x^3$

277. 7.2

278. 10.5 rad/s

279. $v = -20\,\pi\,\sin(2\pi t + \pi/3)$ cm/s

280. 3

281. $R_x = 5$ m; $R_y = 5$ m; $R = 7.07$ m; $\theta_R = 45°$

282. 10^3

283. Zero

284. 4.3°

285. $-2\,\mathbf{v}$ is a velocity of 60 mi/h directed 45° south of due west.

286. 12.30 ± 0.03 km

287. 25°

288. A displacement of 135 m directed 45° north of due west.

289. The displacement x (in meters) increases with time t (in seconds) at a constant rate of 0.25 m/s.

290. $v = \frac{F_0\tau}{m}(1 - e^{-t/\tau})$

292. Zero

293. 10^{-4}

294. 1.72×10^{-2} rad/day

295. $x = 1$ m; $y = -13$ m

296. 0.0401

297. same

298. 0.47 km/s

299. $y = \frac{1}{6}x + \frac{1}{2}$

300. $a = -700\pi^2 \cos(10\pi t)$ cm/s²

301. $-3\,\mathbf{v}$ is a velocity of 90 mi/h directed 45° south of due west.

302. 1

303. $3\,\mathbf{v}$ is a velocity of 90 mi/h directed 45° north of due east.

305. $1 - 3x + 6x^2 - 10x^3 + \cdots$

306. $\tau = 2\alpha mr^2\,\mathbf{k}$

307. $\frac{88}{9}t^{5/3}$

308. $\ll$

309. 6.32 cm

310. 10^3

311. 0.5774

312. 2.72×10^3

313. $+19.6$ m/s

314. $y = \frac{2}{3} x + \frac{23}{3}$

315. 0.5 ± 0.2 m

316. 0.0175

317. 1005 m

318. 7.75 m

319. $3 \tan 3t \sec 3t$

320. 1.05

321. $D = -12$ m

322. 5

323. 2×10^{-3}

324. 0.893

325. $W_{ab} = mgh$

326. 7

327. $\ll$

328. 176.5 ± 0.2 s^{-1}

329. 0.928

330. $1.15°$

331. 0.446

332. $-1/x^2$

333. $nt^{n-1} \tan \alpha t + \alpha t^n \sec^2 \alpha t$

335. 5.20 m

336. $\pm 4.4\% = \pm 17$ m^2

337. $|v_{max}| = 220$ cm/s

338. 0.0025

339. $\mathbf{r}\,(t = 0) \cdot \mathbf{r}\,(t = \frac{3}{2}\,\tau) = -r^2$; After one and one-half periods $(3\tau/2)$, the particle is located on the opposite side of the circle. The negative sign means that $\mathbf{r}\,(t = \frac{3}{2}\,\tau)$ is opposite in direction to $\mathbf{r}\,(t = 0)$.

340. No

341. 0

342. $\cos t$

343. 6.32 cm

344. 6 s

345. Down

346. $v = \frac{8}{m} t$

348. $12 \sin 6t \cos 6t$

349. Rectangle

350. $D = \dfrac{t^{-n+1}}{1 - n}$

351. $\sim 10^{18}$ s

352. $R_x = 0$; $R_y = 330$ m/s; $R = 330$ m/s; $\theta_R = 90°$

353. $T = 1$ s

354. Square

355. 4.8

356. 1.006

357. $\gtrsim$

358. 30 m

359. $x = \frac{2}{3}$, 5

360. $A_x = 86.6$ N
$A_y = 50$ N
$B_x = 37.5$ N
$B_y = 64.95$ N
$R_x = 124.1$ N
$R_y = 114.95$ N
$R = 169.2$ N
$\theta_R = 42.8°$

362. $\mathbf{R} = 8.66 \times 10^3\,\mathbf{i} - 5 \times 10^3\,\mathbf{j}$ (in N)

363. $te^t - e^t + C$

364. $v = \dfrac{F_0}{2m} \sin 2t$

365. $D = 337.5$ cm^2

366. 6752 m/s

367. $3\,t^2$

368. $+\infty$ and $-\infty$

369. $D = \frac{1}{7}\,t^7$

370. $\mathbf{x} = -3.16\,\mathbf{i} - 3.16\,\mathbf{j}$ (in m)

371. $A = 10$ kg·m/s; $\theta = 210°$; $A_x = -8.66$ kg·m/s; $A_y = -5$ kg·m/s

372. Boston Red Sox, with 89 losses and 73 wins

373. $-\dfrac{4}{t^{3/2}} + C$

374. 9.8 m/s^2, 0.28 m/s^2

375. 1

376. $W_{ab} = \frac{1}{4}\,kx_b^4\,(-\frac{1}{2}\,\alpha x_b^2)$

377. $\frac{10}{3}\,t^{29}$

378. $-2 \cos t \sin t$

379. $D = 427.5$ cm^2

380. $a(t) = \dfrac{\alpha}{m} t^2$

  $v(t) = v_0 + \dfrac{\alpha}{3m} t^3$

  $x(t) = x_0 + v_0 t + \dfrac{\alpha}{12m} t^4$

381. 0

382. 5.5

383. $3200\, t^3$ N

384. -0.41

385. $x(t) = r \cos(\alpha t^2)$; $y(t) = r \sin(\alpha t^2)$

Index